Lecture Notes in Mathematics 1840

Editors:
J.–M. Morel, Cachan
F. Takens, Groningen
B. Teissier, Paris

Springer

Berlin
Heidelberg
New York
Hong Kong
London
Milan
Paris
Tokyo

Boris Tsirelson Wendelin Werner

Lectures on Probability Theory and Statistics

Ecole d'Eté de Probabilités
de Saint-Flour XXXII - 2002

Editor: Jean Picard

 Springer

Authors

Boris Tsirelson
School of Mathematics
Tel Aviv University
Tel Aviv 69978
Israel

e-mail: tsirel@tau.ac.il

Wendelin Werner
Laboratoire de Mathématiques
Université Paris-Sud
Bât 425, 91405 Orsay Cedex
France

e-mail: wendelin.werner@math.u-psud.fr

Editor

Jean Picard
Laboratoire de Mathématiques Appliquées
UMR CNRS 6620
Université Blaise Pascal Clermont-Ferrand
63177 Aubière Cedex, France

e-mail: Jean.Picard@math.univ-bpclermont.fr

Cover picture: Blaise Pascal (1623-1662)

Cataloging-in-Publication Data applied for
Bibliographic information published by Die Deutsche Bibliothek

Die Deutsche Bibliothek lists this publication in the Deutsche Nationalbibliografie;
detailed bibliographic data is available in the Internet at http://dnb.ddb.de

Mathematics Subject Classification (2001):
60-01, 60Gxx, 60J65, 60K35, 82B20, 82B27, 82B41

ISSN 0075-8434 Lecture Notes in Mathematics
ISSN 0721-5363 Ecole d'Eté des Probabilités de St. Flour
ISBN 3-540-21316-3 Springer-Verlag Berlin Heidelberg New York

Springer-Verlag is a part of Springer Science + Business Media

springeronline.com

© Springer-Verlag Berlin Heidelberg 2004
Printed in Germany

Typesetting: Camera-ready TeX output by the authors

SPIN: 10994733   41/3142/du - 543210 - Printed on acid-free paper

Preface

Three series of lectures were given at the 32nd Probability Summer School in Saint-Flour (July 7–24, 2002), by Professors Pitman, Tsirelson and Werner. In order to keep the size of the volume not too large, we have decided to split the publication of these courses into two parts. This volume contains the courses of Professors Tsirelson and Werner. The course of Professor Pitman, entitled "Combinatorial stochastic processes", is not yet ready. We thank the authors warmly for their important contribution.

76 participants have attended this school. 33 of them have given a short lecture. The lists of participants and of short lectures are enclosed at the end of the volume.

Finally, we give the numbers of volumes of Springer *Lecture Notes* where previous schools were published.

Lecture Notes in Mathematics

1971: vol 307	1973: vol 390	1974: vol 480	1975: vol 539
1976: vol 598	1977: vol 678	1978: vol 774	1979: vol 876
1980: vol 929	1981: vol 976	1982: vol 1097	1983: vol 1117
1984: vol 1180	1985/86/87: vol 1362	1988: vol 1427	1989: vol 1464
1990: vol 1527	1991: vol 1541	1992: vol 1581	1993: vol 1608
1994: vol 1648	1995: vol 1690	1996: vol 1665	1997: vol 1717
1998: vol 1738	1999: vol 1781	2000: vol 1816	2001: vol 1837

Lecture Notes in Statistics

1986: vol 50	2001: vol 179

Contents

Part I

Boris Tsirelson: Scaling Limit, Noise, Stability

Scaling Limit, Noise, Stability

Boris Tsirelson

School of Mathematics, Tel Aviv University, Tel Aviv 69978, Israel
tsirel@tau.ac.il
//www.tau.ac.il/~tsirel/

Summary. Linear functions of many independent random variables lead to classical noises (white, Poisson, and their combinations) in the scaling limit. Some singular stochastic flows and some models of oriented percolation involve very nonlinear functions and lead to nonclassical noises. Two examples are examined, Warren's 'noise made by a Poisson snake' and the author's 'Brownian web as a black noise'. Classical noises are stable, nonclassical are not. A new framework for the scaling limit is proposed. Old and new results are presented about noises, stability, and spectral measures.

Introduction

Functions of n independent random variables and limiting procedures for $n \to \infty$ are a tenor of probability theory.

Classical limit theorems investigate linear functions, such as $f(\xi_1, \ldots, \xi_n) = (\xi_1 + \cdots + \xi_n)/\sqrt{n}$. The well-known limiting procedure (a classical example of scaling limit) leads to the Brownian motion. Its derivative, the white noise, is not a continuum of independent random variables, but rather an infinitely divisible 'reservoir of independence', a classical example of a continuous product of probability spaces.

Percolation theory investigates some very special nonlinear functions of independent two-valued random variables, either in the limit of an infinite discrete lattice, or in the scaling limit. The latter is now making spectacular progress. The corresponding 'reservoir of independence' is already constructed for oriented percolation (which is much simpler). That is a modern, nonclassical example of a continuous product of probability spaces.

An essential distinction between classical and nonclassical continuous products of probability spaces is revealed by the concept of stability/sensitivity, framed for the discrete case by computer scientists and (in parallel) for the continuous case by probabilists. Everything is stable if and only if the setup is classical.

Some readers prefer discrete models, and treat continuous models as a mean of describing asymptotic behavior. Such readers may skip Sects. 6.2, 6.3, 8.2, 8.3, 8.4. Other readers are interested only in continuous models. They may restrict themselves to Sects. 3.4, 3.5, 4.9, 5.2, 6, 7, 8.

Scaling limit. A new framework for the scaling limit is proposed in Sects. 1.2, 2, 3.1–3.3.

Noise. The idea of a continuous product of probability spaces is formalized by the notions of 'continuous factorization' (Sect. 3.4) and 'noise' (Sect. 3.5). (Some other types of continuous product are considered in [18], [19].) For two nonclassical examples of noise see Sects. 4, 7.

Stability. Stability (and sensitivity) is studied in Sects. 5, 6.1, 6.4. For an interplay between discrete and continuous forms of stability/sensitivity, see especially Sects. 5.3, 6.4.

The spectral theory of noises, presented in Sects. 3.3, 3.4 and used in Sects. 5, 6, generalizes both the Fourier transform on the discrete group $\mathbb{Z}_2^n$ (the Fourier-Walsh transform) and the Itô decomposition into multiple stochastic integrals. For the scaling limit of spectral measures, see Sect. 3.3.

Throughout, either by assumption or by construction, all probability spaces will be Lebesgue-Rokhlin spaces; that is, isomorphic mod 0 to an interval with Lebesgue measure, or a discrete (finite or countable) measure space, or a combination of both.

1 A First Look

1.1 Two Toy Models

The most interesting thing is a scaling limit as a transition from a lattice model to a continuous model. A transition from a finite sequence to an infinite sequence is much simpler, but still nontrivial, as we'll see on simple toy models.

Classical theorems about independent increments are exhaustive, but a small twist may surprise us. I demonstrate the twist on two models, 'discrete' and 'continuous'. The 'continuous' model is a Brownian motion on the circle. The 'discrete' model takes on two values ± 1 only, and increments are treated multiplicatively: $X(t)/X(s)$ instead of the usual $X(t) - X(s)$. Or equivalently, the 'discrete' process takes on its values in the two-element group $\mathbb{Z}_2$; using additive notation we have $\mathbb{Z}_2 = \{0, 1\}$, $1+1 = 0$, increments being $X(t) - X(s)$. In any case, the twist stipulates values in a compact group (the circle, $\mathbb{Z}_2$, etc.), in contrast to the classical theory, where values are in $\mathbb{R}$ (or another linear space). Also, the classical theory assumes continuity (in probability), while our twist does not. The 'continuous' process (in spite of its name) is discontinuous at a single instant $t = 0$. The 'discrete' process is discontinuous at $t = \frac{1}{n}$, $n = 1, 2, \ldots$, and also at $t = 0$; it is constant on $[\frac{1}{n+1}, \frac{1}{n})$ for every n.

Example 1.1. Introduce an infinite sequence of random signs $\tau_1, \tau_2, \ldots$; that is,

$$\mathbb{P}\left(\tau_k = -1\right) = \mathbb{P}\left(\tau_k = +1\right) = \frac{1}{2} \quad \text{for each } k,$$

$$\tau_1, \tau_2, \ldots \quad \text{are independent.}$$

For each n we define a stochastic process $X_n(\cdot)$, driven by $\tau_1, \ldots, \tau_n$, as follows:

a sample path of X_4
(here $\tau_1 = \tau_2 = \tau_4 = -1, \tau_3 = +1$)

$$X_n(t) = \prod_{k:1/n \le 1/k \le t} \tau_k .$$

For $n \to \infty$, finite-dimensional distributions of X_n converge to those of a process $X(\cdot)$. Namely, X consists of countably many random signs, situated on intervals $[\frac{1}{k+1}, \frac{1}{k})$. Almost surely, X has no limit at $0+$. We have

$$\frac{X(t)}{X(s)} = \prod_{k:s<1/k\le t} \tau_k \tag{1.1}$$

whenever $0 < s < t < \infty$. However, (1.1) does not hold when $s < 0 < t$. Here, the product contains infinitely many factors and diverges almost surely;

nevertheless, the increment $X(t)/X(s)$ is well-defined. Each X_n satisfies (1.1) for all s, t (including $s < 0 < t$; of course, $k \leq n$), but X does not. Still, X is an independent increment process (multiplicatively); that is, $X(t_2)/X(t_1), \ldots, X(t_n)/X(t_{n-1})$ are independent whenever $-\infty < t_1 < \cdots < t_n < \infty$. However, we cannot describe the whole X by a countable collection of its independent increments. The infinite sequence of $\tau_k = X(\frac{1}{k}+)/X(\frac{1}{k}-)$ does not suffice since, say, $X(1)$ is independent of $(\tau_1, \tau_2, \ldots)$. Indeed, the global sign change $x(\cdot) \mapsto -x(\cdot)$ is a measure-preserving transformation that leaves all τ_k invariant. The conditional distribution of $X(\cdot)$ given $\tau_1, \tau_2, \ldots$ is concentrated at two functions of opposite global sign. It may seem that we should add to $(\tau_1, \tau_2, \ldots)$ one more random sign τ_∞ independent of $(\tau_1, \tau_2, \ldots)$ such that $X(\frac{1}{k})$ is a measurable function of $\tau_k, \tau_{k+1}, \ldots$ and τ_∞. However, it is impossible. Indeed, $X(1) = \tau_1 \ldots \tau_k X(\frac{1}{k})$. Assuming $X(\frac{1}{k}) = f_k(\tau_k, \tau_{k+1}, \ldots; \tau_\infty)$ we get $f_1(\tau_1, \tau_2, \ldots; \tau_\infty) = \tau_1 \ldots \tau_{k-1} f_k(\tau_k, \tau_{k+1}, \ldots; \tau_\infty)$ for all k. It follows that $f_1(\tau_1, \tau_2, \ldots; \tau_\infty)$ is orthogonal to all functions of the form $g(\tau_1, \ldots, \tau_n) h(\tau_\infty)$ for all n, and thus, to a dense (in L_2) set of functions of $\tau_1, \tau_2, \ldots; \tau_\infty$; a contradiction.

So, for each n the process X_n is driven by (τ_k), but the limiting process X is not.

Example 1.2. (See also [3].) We turn to the other, the 'continuous' model. For any $\varepsilon \in (0, 1)$ we introduce a (complex-valued) stochastic process

$$Y_\varepsilon(t) = \begin{cases} \exp(iB(\ln t) - iB(\ln \varepsilon)) & \text{for } t \geq \varepsilon, \\ 1 & \text{otherwise,} \end{cases}$$

where $B(\cdot)$ is the usual Brownian motion; or rather, $(B(t))_{t \in [0,\infty)}$ and $(B(-t))_{t \in [0,\infty)}$ are two independent copies of the usual Brownian motion. Multiplicative increments $Y_\varepsilon(t_2)/Y_\varepsilon(t_1), \ldots, Y_\varepsilon(t_n)/Y_\varepsilon(t_{n-1})$ are independent whenever $-\infty < t_1 < \cdots < t_n < \infty$, and the distribution of $Y_\varepsilon(t)/Y_\varepsilon(s)$ does not depend on ε as far as $\varepsilon < s < t$ (in fact, the distribution depends on t/s only). The distribution of $Y_\varepsilon(1)$ converges for $\varepsilon \to 0$ to the uniform distribution on the circle $|z| = 1$. The same for each $Y_\varepsilon(t)$. It follows easily that, when $\varepsilon \to 0$, finite dimensional distributions of Y_ε converge to those of some process Y. For every $t > 0$, $Y(t)$ is distributed uniformly on the circle; Y is an independent increment process (multiplicatively), and $Y(t) = 1$ for $t \leq 0$. Almost surely, $Y(\cdot)$ is continuous on $(0, \infty)$, but has no limit at $0+$. We may define $B(\cdot)$ by

$$Y(t) = Y(1) \exp(iB(\ln t)) \quad \text{for } t \in \mathbb{R},$$
$$B(\cdot) \quad \text{is continuous on } \mathbb{R}.$$

Then B is the usual Brownian motion, and

$$\frac{Y(t)}{Y(s)} = \frac{\exp(iB(\ln t))}{\exp(iB(\ln s))} \quad \text{for } 0 < s < t < \infty.$$

However, $Y(1)$ is independent of $B(\cdot)$. Indeed, the global phase change $y(\cdot) \mapsto e^{i\alpha}y(\cdot)$ is a measure preserving transformation that leaves $B(\cdot)$ invariant. The conditional distribution of $Y(\cdot)$ given $B(\cdot)$ is concentrated on a continuum of functions that differ by a global phase (distributed uniformly on the circle). Similarly to the 'discrete' example, we cannot introduce a random variable $B(-\infty)$ independent of $B(\cdot)$, such that $Y(t)$ is a function of $B(-\infty)$ and increments of $B(r)$ for $-\infty < r < \ln t$.

So, for each ε, the process Y_ε is driven by the Brownian motion, but the limiting process Y is not.

Both toy models are singular at a given instant $t = 0$. Interestingly, continuous stationary processes can demonstrate such strange behavior, distributed in time! (See Sects. 4, 7).

1.2 Our Limiting Procedures

Imagine a sequence of elementary probabilistic models such that the n-th model is driven by a finite sequence $(\tau_1, \ldots, \tau_n)$ of random signs (independent, as before). A limiting procedure may lead to a model driven by an infinite sequence $(\tau_1, \tau_2, \ldots)$ of random signs. However, it may also lead to something else, as shown in Sect. 1.1. This is an opportunity to ask ourselves: what do we mean by a limiting procedure?

The n-th model is naturally described by the finite probability space $\Omega_n = \{-1, +1\}^n$ with the uniform measure. A prerequisite to any limiting procedure is some structure able to join these Ω_n somehow. It may be a sequence of 'observables', that is, functions on the disjoint union,

$$f_k : (\Omega_1 \uplus \Omega_2 \uplus \ldots) \to \mathbb{R}.$$

Example 1.3. Let $f_k(\tau_1, \ldots, \tau_n) = \tau_k$ for $n \geq k$. Though f_k is defined only on $\Omega_k \uplus \Omega_{k+1} \uplus \ldots$, it is enough. For every k, the joint distribution of $f_1, \ldots, f_k$ on Ω_n has a limit for $n \to \infty$ (moreover, the distribution does not depend on n, as far as $n \geq k$). The limiting procedure should extend each f_k to a new probability space Ω such that the joint distribution of $f_1, \ldots, f_k$ on Ω_n converges for $n \to \infty$ to their joint distribution on Ω. Clearly, we may take the space of infinite sequences $\Omega = \{-1, +1\}^\infty$ with the product measure, and let f_k be the k-th coordinate function.

Example 1.4. Still $f_k(\tau_1, \ldots, \tau_n) = \tau_k$ (for $n \geq k \geq 1$), but in addition, the product $f_0(\tau_1, \ldots, \tau_n) = \tau_1 \ldots \tau_n$ is included. For every k, the joint distribution of $f_0, f_1, \ldots, f_k$ on Ω_n has a limit for $n \to \infty$; in fact, the distribution does not depend on n, as far as $n > k$ (this time, not just $n \geq k$). Thus, in the limit, $f_0, f_1, f_2, \ldots$ become independent random signs. The functional dependence $f_0 = f_1 f_2 \ldots$ holds for each n, but disappears in the limit. We still may take $\Omega = \{-1, +1\}^\infty$, however, f_0 becomes a new coordinate.

This is instructive; the limiting model depends on the class of 'observables'.

Example 1.5. Let $f_k(\tau_1, \ldots, \tau_n) = \tau_k \ldots \tau_n$ for $n \geq k \geq 1$. In the limit, f_k become independent random signs. We may define τ_k in the limiting model by $\tau_k = f_k/f_{k+1}$; however, we cannot express f_k in terms of τ_k. Clearly, it is the same as the 'discrete' toy model of Sect. 1.1.

The second and third examples are isomorphic. Indeed, renaming f_k of the third example as g_k (and retaining f_k of the second example) we have

$$g_k = \frac{f_0}{f_1 \ldots f_{k-1}}; \qquad f_k = \frac{g_k}{g_{k+1}} \text{ for } k > 0, \quad \text{and} \quad f_0 = g_1;$$

these relations hold for every n (provided that the same $\Omega_n = \{-1, +1\}^n$ is used for both examples) and naturally, give us an isomorphism between the two limiting models.

That is also instructive; some changes of the class of 'observables' are essential, some are not.

It means that the sequence (f_k) is not really the structure responsible for the limiting procedure. Rather, f_k are generators of the relevant structure. The second and third examples differ only by the choice of generators for the same structure. In contrast, the first example uses a different structure. So, what is the mysterious structure?

I can describe the structure in two equivalent ways. Here is the first description. In the commutative Banach algebra $l_\infty(\Omega_1 \uplus \Omega_2 \uplus \ldots)$ of all bounded functions on the disjoint union, we select a subset C (its elements will be called observables) such that

C is a separable closed subalgebra of $l_\infty(\Omega_1 \uplus \Omega_2 \uplus \ldots)$ containing the unit.

$$(1.2)$$

In other words,

$$C \text{ contains a sequence dense in the uniform topology;}$$
$$f_n \in C, \ f_n \to f \text{ uniformly} \quad \Longrightarrow \quad f \in C;$$
$$f, g \in C, \ a, b \in \mathbb{R} \quad \Longrightarrow \quad af + bg \in C; \qquad (1.3)$$
$$\mathbf{1} \in C;$$
$$f, g \in C \quad \Longrightarrow \quad fg \in C$$

(here $\mathbf{1}$ stands for the unity, $\mathbf{1}(\omega) = 1$ for all ω). Or equivalently,

$$C \text{ contains a sequence dense in the uniform topology;}$$
$$f_n \in C, \ f_n \to f \text{ uniformly} \quad \Longrightarrow \quad f \in C; \qquad (1.4)$$
$$f, g \in C, \ \varphi : \mathbb{R}^2 \to \mathbb{R} \text{ continuous} \quad \Longrightarrow \quad \varphi(f, g) \in C.$$

Indeed, on one hand, both $af + bg$ and fg (and $\mathbf{1}$) are special cases of $\varphi(f, g)$. On the other hand, every continuous function on a bounded subset of $\mathbb{R}^2$ can be uniformly approximated by polynomials. The same holds for $\varphi(f_1, \ldots, f_n)$

where $f_1, \ldots, f_n \in C$, and $\varphi : \mathbb{R}^n \to \mathbb{R}$ is a continuous function. Another equivalent set of conditions is also well-known:

$$C \text{ contains a sequence dense in the uniform topology;}$$
$$f_n \in C, \; f_n \to f \text{ uniformly} \implies f \in C;$$
$$f, g \in C, \; a, b \in \mathbb{R} \implies af + bg \in C; \tag{1.5}$$
$$1 \in C;$$
$$f \in C \implies |f| \in C;$$

here $|f|$ is the pointwise absolute value, $|f|(\omega) = |f(\omega)|$.

The smallest set C satisfying these (equivalent) conditions (1.2)–(1.5) and containing all given functions f_k is, by definition, generated by these f_k.

Recall that C consists of functions defined on the disjoint union of finite probability spaces Ω_n; a probability measure P_n is given on each Ω_n. The following condition is relevant:

$$\lim_{n \to \infty} \int_{\Omega_n} f \, dP_n \text{ exists for every } f \in C. \tag{1.6}$$

Assume that C is generated by given functions f_k. Then the property (1.6) of C is equivalent to such a property of functions f_k:

For each k, the joint distribution of $f_1, \ldots, f_k$ on Ω_n weakly converges, when $n \to \infty$. $\tag{1.7}$

Proof: (1.7) means convergence of $\int \varphi(f_1, \ldots, f_k) \, dP_n$ for every continuous function $\varphi : \mathbb{R}^k \to \mathbb{R}$. However, functions of the form $f = \varphi(f_1, \ldots, f_k)$ (for all k, φ) belong to C and are dense in C.

We see that (1.7) does not depend on the choice of generators f_k of a given C.

The second (equivalent) description of our structure is the 'joint compactification' of $\Omega_1, \Omega_2, \ldots$ I mean a pair (K, α) such that

$$K \text{ is a metrizable compact topological space,}$$
$$\alpha : (\Omega_1 \uplus \Omega_2 \uplus \ldots) \to K \text{ is a map,} \tag{1.8}$$
$$\text{the image } \alpha(\Omega_1 \uplus \Omega_2 \uplus \ldots) \text{ is dense in } K.$$

Every joint compactification (K, α) determines a set C satisfying (1.2). Namely,

$$C = \alpha^{-1}\big(C(K)\big);$$

that is, observables $f \in C$ are, by definition, functions of the form

$$f = g \circ \alpha, \text{ that is, } f(\omega) = g(\alpha(\omega)), \quad g \in C(K).$$

The Banach algebra C is basically the same as the Banach algebra $C(K)$ of all continuous functions on K.

Every C satisfying (1.2) corresponds to some joint compactification. Proof: C is generated by some f_k such that $|f_k(\omega)| \le 1$ for all k, ω. We introduce

$$\alpha(\omega) = \big(f_1(\omega), f_2(\omega), \dots\big) \in [-1, 1]^\infty,$$
$$K \text{ is the closure of } \alpha(\Omega_1 \uplus \Omega_2 \uplus \dots) \text{ in } [-1, 1]^\infty;$$

clearly, (K, α) is a joint compactification. Coordinate functions on K generate $C(K)$, therefore f_k generate $\alpha^{-1}\big(C(K)\big)$, hence $\alpha^{-1}\big(C(K)\big) = C$.

Finiteness of each Ω_n is not essential. The same holds for arbitrary probability spaces $(\Omega_n, \mathcal{F}_n, P_n)$. Of course, instead of $l_\infty(\Omega_1 \uplus \Omega_2 \uplus \dots)$ we use $L_\infty(\Omega_1 \uplus \Omega_2 \uplus \dots)$, and the map $\alpha : (\Omega_1 \uplus \Omega_2 \uplus \dots) \to K$ must be measurable. It sends the given measure P_n on Ω_n into a measure $\alpha(P_n)$ (denoted also by $P_n \circ \alpha^{-1}$) on K. If measures $\alpha(P_n)$ weakly converge, we get the limiting model (Ω, P) by taking $\Omega = K$ and $P = \lim_{n \to \infty} \alpha(P_n)$.

1.3 Examples of High Symmetry

Example 1.6. Let Ω_n be the set of all permutations $\omega : \{1, \dots, n\} \to \{1, \dots, n\}$, each permutation having the same probability $(1/n!)$;

$$f : (\Omega_1 \uplus \Omega_2 \uplus \dots) \to \mathbb{R} \text{ is defined by}$$
$$f(\omega) = |\{k : \omega(k) = k\}|;$$

that is, the number of fixed points of a random permutation. Though f is not bounded, which happens quite often, in order to embed it into the framework of Sect. 1.2, we make it bounded by some homeomorphism from $\mathbb{R}$ to a bounded interval (say, $\omega \mapsto \arctan f(\omega)$). The distribution of $f(\cdot)$ on Ω_n converges (for $n \to \infty$) to the Poisson distribution $P(1)$. Thus, the limiting model exists; however, it is scanty: just $P(1)$.

We may enrich the model by introducing

$$f_u(\omega) = |\{k < un : \omega(k) = k\}|;$$

for instance, $f_{0.5}(\cdot)$ is the number of fixed points among the first half of $\{1, \dots, n\}$. The parameter u could run over $[0, 1]$, but we need a countable set of functions; thus we restrict u to, say, rational points of $[0, 1]$. Now the limiting model is the Poisson process.

Each finite model here is invariant under permutations. Functions f_u seem to break the invariance, but the latter survives in their increments, and turns in the limit into invariance of the Poisson process (or rather, its derivative, the point process) under all measure preserving transformations of $[0, 1]$.

Note also that *independent* increments in the limit emerge from *dependent* increments in finite models.

We feel that all these $f_u(\cdot)$ catch only a small part of the information contained in the permutation. You may think about more information, say, cycles of length $1, 2, \dots$ (and what about length $n/2$?)

Example 1.7. Let Ω_n be the set of all graphs over $\{1, \ldots, n\}$. That is, each $\omega \in \Omega_n$ is a subset of the set $\binom{\{1, \cdots, n\}}{2}$ of all unordered pairs (treated as edges, while $1, \ldots, n$ are vertices); the probability of ω is $p_n^{|\omega|}(1 - p_n)^{n(n-1)/2 - |\omega|}$, where $|\omega|$ is the number of edges. That is, every edge is present with probability p_n, independently of others. Define $f(\omega)$ as the number of isolated vertices. The limiting model exists if (and only if) there exists a limit $\lim_n n(1 - p_n)^{n-1} = \lambda \in [0, \infty)$;[1] the Poisson distribution $P(\lambda)$ exhausts the limiting model.

A Poisson process may be obtained in the same way as before.

You may also count small connected components which are more complicated than single points.

Note that the finite model contains a lot of independence (namely, $n(n - 1)/2$ independent random variables); the limiting model (Poisson process) also contains a lot of independence (namely, independent increments). However, we feel that independence is not inherited; rather, the independence of finite models is lost in the limiting procedure, and a new independence emerges.

Example 1.8. Let $\Omega_n = \{-1, +1\}^n$ with uniform measure, and $f_n : (\Omega_1 \uplus \Omega_2 \uplus \ldots) \to \mathbb{R}$ be defined by

$$f_u(\omega) = \frac{1}{\sqrt{n}} \sum_{k < un} \tau_k(\omega) \, ;$$

as before, $\tau_1, \ldots, \tau_n$ are the coordinates, that is, $\omega = (\tau_1(\omega), \ldots, \tau_n(\omega))$ and u runs over rational points of $[0, 1]$. The limiting model is the Brownian motion, of course.

Similarly to Example 1.6, each finite model is invariant under permutations. The invariance survives in increments of functions f_k, and in the limit, the white noise (the derivative of the Brownian motion) is invariant under all measure preserving transformations of $[0, 1]$.

A general argument of Sect. 6.3 will show that a high symmetry model cannot lead to a nonclassical scaling limit.

1.4 Example of Low Symmetry

Example 1.8 may be rewritten via the composition of random maps

$$\alpha_-, \alpha_+ : \mathbb{Z} \to \mathbb{Z} \, ,$$
$$\alpha_-(k) = k - 1, \quad \alpha_+(k) = k + 1 \, ;$$
$$\alpha_\omega = \alpha_{\tau_n(\omega)} \circ \ldots \alpha_{\tau_1(\omega)} \, ;$$

α_- α_+

---

[1] Formally, the limiting model exists also for $\lambda = \infty$, since the range of f is compactified.

thus, $\alpha_\omega(k) = k + \tau_1(\omega) + \cdots + \tau_n(\omega)$, and we may define $f_1(\omega) = \frac{1}{\sqrt{n}}\alpha_\omega(0)$, which conforms to Example 1.8. Similarly, $f_u(\omega) = \frac{1}{\sqrt{n}}\alpha_{\omega,u}(0)$, where $\alpha_{\omega,u}$ is the composition of $\alpha_{\tau_k(\omega)}$ for $k \leq un$. The order does not matter, since α_-, α_+ commute, that is, $\alpha_- \circ \alpha_+ = \alpha_+ \circ \alpha_-$. It is interesting to try a pair of noncommuting maps.

Example 1.9. (See Warren [22].) Define

$$\alpha_-, \alpha_+ : \mathbb{Z} + \frac{1}{2} \to \mathbb{Z} + \frac{1}{2},$$

$$\alpha_-(x) = x - 1,$$
$$\alpha_+(x) = x + 1 \quad \text{for } x \in \left(\mathbb{Z} + \tfrac{1}{2}\right) \cap (0, \infty),$$

$$\alpha_-(-x) = -\alpha_-(x), \quad \alpha_+(-x) = -\alpha_+(x).$$

These are not invertible functions; α_- is not injective, α_+ is not surjective. Well, we do not need to invert them, but need their compositions:

$$\alpha_\omega = \alpha_{\tau_n(\omega)} \circ \cdots \circ \alpha_{\tau_1(\omega)}.$$

$$\alpha_- \circ \alpha_+ = \alpha_{(+1,-1)} \qquad \alpha_+ \circ \alpha_- = \alpha_{(-1,+1)}$$

All compositions belong to a two-parameter set of functions $h_{a,b}$,

$$\alpha_\omega(x) = h_{a,b}(x) = \begin{cases} x + a & \text{for } x \geq b, \\ x - a & \text{for } x \leq -b, \\ (-1)^{b-x}(a+b) & \text{for } -b \leq x \leq b; \end{cases}$$

$$y = \alpha_\omega(x)$$

$$b, a + b \in \left(\mathbb{Z} + \tfrac{1}{2}\right) \cap (0, \infty) = \left\{\tfrac{1}{2}, \tfrac{3}{2}, \tfrac{5}{2}, \dots\right\}.$$

Indeed, $\alpha_- = h_{-1,1.5}$, $\alpha_+ = h_{1,0.5}$, and $h_{a_2,b_2} \circ h_{a_1,b_1} = h_{a,b}$ where $a = a_1 + a_2$, $b = \max(b_1, b_2 - a_1)$. Thus, $\alpha_\omega = h_{a(\omega),b(\omega)}$, and we define

$$f_1 : (\Omega_1 \uplus \Omega_2 \uplus \dots) \to \mathbb{R}^2 \times \{-1, +1\},$$

$$f_1(\omega) = \left(\frac{a(\omega)}{\sqrt{n}}, \frac{b(\omega)}{\sqrt{n}}, (-1)^{b(\omega)-0.5}\right).$$

However, the function is neither bounded nor real-valued; in order to fit into the framework of Sect. 1.2 we take, say, $\arctan(a(\omega)/\sqrt{n})$, $\arctan(b(\omega)/\sqrt{n})$, and $(-1)^{b(\omega)-0.5}$. The latter is essential if, say, $\frac{1}{\sqrt{n}}\alpha_\omega(0.5)$ is treated as an 'observable'; indeed, $\frac{1}{\sqrt{n}}\alpha_\omega(0.5) = (-1)^{b(\omega)-0.5}\frac{1}{\sqrt{n}}(a(\omega) + b(\omega))$. The limiting model exists, and is quite interesting. (See also Sect. 8.3.) As before, a random process appears by considering the composition over $k < un$.

Here, finite models are not invariant under permutations of their independent random variables (since the maps do not commute), and the limiting model appears not to be invariant under measure preserving transformations of $[0, 1]$.

Independence present in finite models survives in the limit, provided that the limit is described by a two-parameter random process; we'll return to this point in Sect. 4.3.

1.5 Trees, Not Cubes

Example 1.10. A particle moves on the sphere S^2. Initially it is at a given point $x_0 \in S^2$. Then it jumps by ε in a random direction. That is, $X_0 = x_0$, while the next random variable X_1 is distributed uniformly on the circle $\{x \in S^2 : |x_0 - x| = \varepsilon\}$. Then it jumps again to X_2 such that $|X_1 - X_2| = \varepsilon$, and so on. We have a Markov chain (X_k) in discrete time (and continuous space). Let Ω_ε be the corresponding probability space; it may be the space of sequences $(x_0, x_1, x_2, \dots)$ satisfying $|x_k - x_{k+1}| = \varepsilon$, or something else, but in any case $X_k : \Omega_\varepsilon \to S^2$. We choose $\varepsilon_n \to 0$ (say, $\varepsilon_n = 1/n$), take $\Omega_n = \Omega_{\varepsilon_n}$ and define $f_u : (\Omega_1 \uplus \Omega_2 \uplus \dots) \to S^2$ by

$$f_u(\omega) = X_k(\omega) \quad \text{for } \varepsilon_n^2 k \le u < \varepsilon_n^2 (k+1), \quad \omega \in \Omega_n.$$

Of course, the limiting model is the Brownian motion on the sphere S^2.

In contrast to previous examples, here Ω_n is not a product; the n-th model does not consist of independent random variables. But, though we can parameterize these Markov transitions by independent random variables, there is a lot of freedom in doing so; none of the parameterizations may be called canonical. The same holds for the limiting model. The Brownian motion on S^2 can be driven by the Brownian motion on R^2 according to some stochastic differential equation, but the latter involves a lot of freedom.

Example 1.11. (See [12].) Consider the random walk on such an oriented graph:

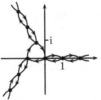

A particle starts at 0 and chooses at random (with probabilities $1/2$, $1/2$) one of the two outgoing edges, and so on (you see, exactly two edges go out of any vertex). Such $(Z_0, Z_1, \dots)$ is known as the simplest spider walk. It is a complex-valued martingale. The set Ω_n of all n-step trajectories contains 2^n elements and carries its natural structure of a binary tree. (It can be mapped to the binary cube $\{-1, +1\}^n$ in many ways.) We define $f_u : (\Omega_1 \uplus \Omega_2 \uplus \ldots) \to \mathbb{C}$ by

$$f_u(\omega) = \frac{1}{\sqrt{n}} Z_k(\omega) \quad \text{for } k \leq nu < k+1, \quad \omega \in \Omega_n.$$

The limiting model is a continuous complex-valued martingale whose values belong to the union of three rays.

The process is known as Walsh's Brownian motion, a special case of the so-called spider martingale.

1.6 Sub-σ-fields

Every example considered till now follows the pattern of Sect. 1.2; a joint compactification of probability spaces Ω_n, and the limiting Ω. Moreover, Ω_n is usually related to a set T_n (a parameter space, interpreted as time or space), and Ω to a joint compactification T of these T_n.

Example	T_n	T
1.1	$\{1, \frac{1}{2}, \ldots, \frac{1}{n}\}$	$\{1, \frac{1}{2}, \frac{1}{3}, \ldots\} \cup \{0\}$
1.2	$[\varepsilon_n, 1]$	$[0, 1]$
1.6, 1.7, 1.8, 1.9, 1.10, 1.11	$\{\frac{1}{n}, \frac{2}{n}, \ldots, 1\}$	$[0, 1]$

Examples 1.1, 1.2, 1.8 deal (for a finite n) with independent increment processes, taking on their values in a group, namely, 1.8: $\mathbb{R}$ (additive); 1.1: $\{-1, +1\}$ (multiplicative), 1.2: the circle $\{z \in \mathbb{C} : |z| = 1\}$ (multiplicative). Every $t \in T_n$ splits the process into two parts, the past and the future; in order to keep them independent, we define them via increments, not values.[2] In terms of random signs τ_k (for 1.1, 1.8) it means simply $\{-1, +1\}^n = \{-1, +1\}^k \times \{-1, +1\}^{n-k}$; here k depends on t. The same idea (of independent parts) is formalized by sub-σ-fields $\mathcal{F}_{0,t}$ (the past) and $\mathcal{F}_{t,1}$ (the future) on our probability space (Ω_m or Ω). Say, for the Brownian motion 1.8, $\mathcal{F}_{0,t}$ is generated by Brownian increments on $[0, t]$, while $\mathcal{F}_{t,1}$ — on $[t, 1]$. Similarly we may define $\mathcal{F}_{s,t}$ for $s < t$, and we have

$$\mathcal{F}_{r,s} \otimes \mathcal{F}_{s,t} = \mathcal{F}_{r,t} \quad \text{whenever } r < s < t.$$

It means two things: first, independence,

---

[2] In fact, the process of Example 1.1 has also independent values (not only increments); but that is irrelevant.

$$\mathbb{P}\left(A\cap B\right) = \mathbb{P}\left(A\right)\mathbb{P}\left(B\right) \quad \text{whenever } A \in \mathcal{F}_{r,s}, B \in \mathcal{F}_{s,t};$$

and second, $\mathcal{F}_{r,t}$ is generated by $\mathcal{F}_{r,s}$ and $\mathcal{F}_{s,t}$ (that is, $\mathcal{F}_{r,t}$ is the least sub-σ-field containing both $\mathcal{F}_{r,s}$ and $\mathcal{F}_{s,t}$). Such a two-parameter family $(\mathcal{F}_{s,t})$ of sub-σ-fields is called a *factorization* (of the given probability space). Some additional precautions are needed when dealing with semigroups (like Example 1.9), and also, with discrete time.

Sub-σ-fields $\mathcal{F}_A$ can be defined for some subsets $A \subset T$ more general than intervals, getting

$$\mathcal{F}_A \otimes \mathcal{F}_B = \mathcal{F}_C \quad \text{whenever } A \uplus B = C.$$

Models of high symmetry admit arbitrary measurable sets A; models of low symmetry do not. For some examples (such as 1.6, 1.7), a factorization emerges after the limiting procedure.[3]

No factorization at all is given for 1.10, 1.11. Still, the past $\mathcal{F}_{0,t} = \mathcal{F}_t$ is defined naturally. However, the future is not defined, since possible continuations depend on the past. Here we deal with a one-parameter family $(\mathcal{F}_t)$ of sub-σ-fields, satisfying only a monotonicity condition

$$\mathcal{F}_s \subset \mathcal{F}_t \quad \text{whenever } s < t;$$

such $(\mathcal{F}_t)$ is called a *filtration*.

---

[3] For Example 1.7, some factorization is naturally defined for Ω_n, but is lost in the limiting procedure, and a new factorization emerges.

2 Abstract Nonsense of the Scaling Limit

2.1 More on Our Limiting Procedures

The joint compactification K of $\Omega_1 \uplus \Omega_2 \uplus \dots$, used in Sect. 1.2, is not quite satisfactory. Return to Example 1.8:

$$f_u(\omega) = \frac{1}{\sqrt{n}} \sum_{k<un} \tau_k(\omega) \quad \text{for } u \in [0,1] \cap \mathbb{Q} \tag{2.1}$$

($\mathbb{Q}$ being the set of rational numbers). The limiting model is the Brownian motion, restricted to $[0,1] \cap \mathbb{Q}$. What about an irrational point, $v \in [0,1] \setminus \mathbb{Q}$? The random variable f_v may be defined on Ω as the limit (say, in L_2) of f_u for $u \to v$, $u \in [0,1] \cap \mathbb{Q}$. On the other hand, f_v is naturally defined on $\Omega_1 \uplus \Omega_2 \uplus \dots$ (by the same formula (2.1)). However, f_v is not a continuous function on the compact space K.[4] Thus, the weak convergence $P_i \to P$ is relevant to f_u but not f_v. Something is wrong!

What is wrong is the uniform topology used in (1.2)–(1.5). A right topology should take measures P_i into account. We have two ways, 'moderate' and 'radical'.

Here is the 'moderate' way. We choose some appropriate subsets $B_m \subset (\Omega_1 \uplus \Omega_2 \uplus \dots)$, $B_1 \subset B_2 \subset \dots$, such that

$$\inf_i P_i(B_m \cap \Omega_i) \uparrow 1 \quad \text{for } m \to \infty$$

and in (1.3)–(1.5) replace the assumption "$f_n \in C$, $f_n \to f$ uniformly $\Longrightarrow$ $f \in C$" with

$$f_n \in C, \ f_n \to f \text{ uniformly on each } B_m \quad \Longrightarrow \quad f \in C. \tag{2.2}$$

Example 2.1. Continuing (2.1) we define B_m by

$$B_m \cap \Omega_i = \left\{ \omega \in \Omega_i : \ \sup_{0 \le k < l \le i} \frac{\left| \frac{1}{\sqrt{i}} \sum_{j=k}^l \tau_j(\omega) \right|}{\left(\frac{l-k}{i} \right)^{1/3}} \le m \right\} ;$$

then[5]

$$|f_u(\omega) - f_v(\omega)| \le m|u-v|^{1/3} \quad \text{for } \omega \in B_m \cap \Omega_i$$

if i is large enough (namely, $2/i < |u-v|$). The set C (satisfying (2.2)) generated by f_u for all rational u, also contains f_v for all irrational v.

---

[4] There exist $\omega_n \in \Omega_n$ such that $\lim_n f_u(\omega_n)$ exists for all $u \in [0,1] \cap \mathbb{Q}$, but $\lim_n f_v(\omega_n)$ does not exist.

[5] Of course, $|u-v|^\alpha$ for any $\alpha \in (0,1/2)$ may be used, not only $|u-v|^{1/3}$.

Similarly to Sect. 1.2, we may translate (2.2) into the topological language. For each m, the restriction of C to B_m corresponds to a joint compactification (K_m, α_m) of $B_m \cap \Omega_i$. Clearly, $K_{m_1} \subset K_{m_2}$ for $m_1 < m_2$, and $\alpha_{m_1} = \alpha_{m_2}|_{K_{m_1}}$. Thus, we get a *joint σ-compactification*

$$\alpha : (\Omega_1 \uplus \Omega_2 \uplus \ldots) \to K_\infty = K_1 \cup K_2 \cup \ldots$$

We do not need a topology on the union K_∞ of metrizable compact spaces $K_1 \subset K_2 \subset \ldots$ [6] We just define $C(K_\infty)$ as the set of all functions $g : K_\infty \to \mathbb{R}$ such that $g|_{K_m}$ is continuous (on K_m) for each m. We have

$$C = \alpha^{-1}\big(C(K_\infty)\big),$$

that is, observables $f \in C$ are functions of the form

$$f = g \circ \alpha, \quad \text{that is, } f(\omega) = g(\alpha(\omega)), \quad g \in C(K_\infty).$$

If measures $\alpha(P_i)$ weakly converge (w.r.t. bounded functions of $C(K_\infty)$, recall (1.6), (1.7)), we get the limiting model (Ω, P) by taking $\Omega = K_\infty$ and $P = \lim_{i \to \infty} \alpha(P_i)$.

Example 2.2. Continuing Example 2.1 we see that the limiting measure P exists, and the joint distribution of all f_u (extended to K_∞ by continuity) w.r.t. P is the Wiener measure. The 'uniform' metric on K_∞,

$$\text{dist}(x, y) = \sup_{0 \le u \le 1} |f_u(x) - f_u(y)|,$$

is continuous on each K_m. Therefore, every function continuous in the 'uniform' metric belongs to $C(K_\infty)$. Our joint σ-compactification is another form of the usual weak convergence of random walks to the Brownian motion.

That was the 'moderate way'. It requires special subsets $B_m \subset (\Omega_1 \uplus \Omega_2 \uplus \ldots)$, in contrast to the 'radical way'; basically, the latter allows the sequence of sets B_m to depend on a sequence of functions f_n, see (2.2). In other words, instead of uniform (or 'locally uniform') convergence, we introduce a weaker topology by the metric[7]

$$\text{dist}(f, g) = \sup_i \int \frac{|f(\omega) - g(\omega)|}{1 + |f(\omega) - g(\omega)|} \, dP_i(\omega). \tag{2.3}$$

---

[6] But if you want, K_∞ may be equipped with the inductive limit topology; that is, $U \subset K_\infty$ is open if and only if for every m, $U \cap K_m$ is open (in K_m). However, the topology usually is not metrizable.

[7] Alternatively, we may restrict ourselves to bounded functions $\Omega_1 \uplus \Omega_2 \uplus \cdots \to [-1, +1]$ (applying a transformation like arctan) and use, say,

$$\text{dist}(f, g) = \sup_i \int |f(\omega) - g(\omega)| \, dP_i(\omega).$$

If $f_n \in C(K)$ and $\mathrm{dist}(f_n, f) \to 0$ then f_n converge in probability w.r.t. P; thus, f is naturally defined P-almost everywhere.[8]

Let C be the closure of $C(K)$ in the metric (2.3). Then

$$\int \varphi(f_1, \ldots, f_d) \, \mathrm{d}P_i \xrightarrow[i \to \infty]{} \int \varphi(f_1, \ldots, f_d) \, \mathrm{d}P$$

for every d, every bounded continuous function $\varphi : \mathbb{R}^d \to \mathbb{R}$, and every $f_1, \ldots, f_d \in C$. The joint distribution of $f_1, \ldots, f_d$ w.r.t. P_i converges (weakly) to that w.r.t. P. So, the weak convergence $P_i \to P$ is relevant for the whole C (not only $C(K)$). That is the idea of the 'radical way', presented systematically in Sects. 2.2, 2.3.

Returning again to Example 1.8 we see that f_v (for $v \in [0, 1]$) is the limit of f_u (for $u \in [0, 1] \cap \mathbb{Q}$) in the metric (2.3); thus, $f_v \in C$ for all $v \in [0, 1]$.

However, much more can be said. Not only

$$\mathrm{Lim}_{i \to \infty} \left(\frac{1}{\sqrt{i}} \sum_{ai < k < bi} \tau_k(\omega) \right) = \int_a^b \mathrm{d}B(t),$$

where 'Lim' means the scaling limit (as explained above), but also

$$\mathrm{Lim}_{i \to \infty} \left(i^{-d/2} \sum_{ai < k_1 < \cdots < k_d < bi} \tau_{k_1}(\omega) \ldots \tau_{k_d}(\omega) \right)$$

$$= \int \cdots \int_{a < t_1 < \cdots < t_d < b} \mathrm{d}B(t_1) \ldots \mathrm{d}B(t_d) = \frac{1}{d!} H_d \big(B(b) - B(a), b - a \big)$$

where H_d is the Hermite polynomial (see for instance [11, IV.3.8]). Taking finite linear combinations and their closure in the metric (2.3) we get

$$\mathrm{Lim}_{i \to \infty} \left(\sum_{d=0}^{\infty} i^{-d/2} \sum_{0 < k_1 < \cdots < k_d < i} \psi_d\left(\tfrac{k_1}{i}, \ldots, \tfrac{k_d}{i}\right) \tau_{k_1}(\omega) \ldots \tau_{k_d}(\omega) \right)$$

$$= \sum_{d=0}^{\infty} \int \cdots \int_{0 < t_1 < \cdots < t_d < 1} \psi_d(t_1, \ldots, t_d) \, \mathrm{d}B(t_1) \ldots \mathrm{d}B(t_d) \quad (2.4)$$

provided that functions ψ_d are Riemann integrable, and vanish for d large enough. The right-hand side is well-defined for all $\psi_d \in L_2$ such that $\sum_d \|\psi_d\|_2^2 < \infty$; the scaling limit may be kept by replacing $\psi_d\left(\tfrac{k_1}{i}, \ldots, \tfrac{k_d}{i}\right)$ with the mean value of ψ_d on the $1/i$-cube centered at $\left(\tfrac{k_1}{i}, \ldots, \tfrac{k_d}{i}\right)$. Now, $(0, 1)$

---

[8] In fact, every (equivalence class of) P-measurable function can be obtained in that way provided that, for each i, supports of P_i and P do not intersect. It means that every random variable on the limiting probability space is the scaling limit of some function on $\Omega_1 \uplus \Omega_2 \uplus \ldots$ (see also Remark 2.15).

may be replaced with the whole $\mathbb{R}$; ψ_d is defined on $\Delta_d = \{(x_1, \ldots, x_d) \in \mathbb{R}^d : x_1 < \cdots < x_d\}$. The right-hand side of (2.4) gives us an isometric linear correspondence between $L_2(\Delta_0 \uplus \Delta_1 \uplus \Delta_2 \uplus \ldots)$ and $L_2(\Omega, \mathcal{F}, P)$, where $(\Omega, \mathcal{F}, P)$ is the probability space describing the Brownian motion (on the whole $\mathbb{R}$).

2.2 Coarse Probability Space: Definition and Simple Example

Definition 2.3. *A* coarse probability space $\left((\Omega[i], \mathcal{F}[i], P[i])_{i=1}^{\infty}, \mathcal{A}\right)$ *consists of a sequence of probability spaces* $(\Omega[i], \mathcal{F}[i], P[i])$ *and a set* $\mathcal{A}$ *of subsets of the disjoint union* $\Omega[\text{all}] = \Omega(1) \uplus \Omega(2) \uplus \ldots$, *satisfying the following conditions:*

(a) $\forall A \in \mathcal{A} \; \forall i \; (A \cap \Omega[i]) \in \mathcal{F}[i]$;

(b) $\forall A, B \in \mathcal{A} \; \left(A \cap B \in \mathcal{A}, \; A \cup B \in \mathcal{A}, \; \Omega[\text{all}] \setminus A \in \mathcal{A}\right)$;

(c) $\mathcal{A}$ *contains every* $A \subset \Omega[\text{all}]$ *such that* $\forall i \; (A \cap \Omega[i]) \in \mathcal{F}[i]$ *and* $P[i]\left(A \cap \Omega[i]\right) \to 0$ *for* $i \to \infty$;

(d) $\left(\cup_{k=1}^{\infty} A_k\right) \in \mathcal{A}$ *for every pairwise disjoint* $A_1, A_2, \cdots \in \mathcal{A}$ *such that* $\sum_k \sup_i P[i]\left(A_k \cap \Omega[i]\right) < \infty$;

(e) $\lim_i P[i]\left(A \cap \Omega[i]\right)$ *exists for every* $A \in \mathcal{A}$;

(f) *there exists a finite or countable subset* $\mathcal{A}_1 \subset \mathcal{A}$ *that generates* $\mathcal{A}$ *in the sense that the least subset of* $\mathcal{A}$ *satisfying* (b)–(d) *and containing* $\mathcal{A}_1$ *is the whole* $\mathcal{A}$.

A set $\mathcal{A}$ *satisfying* (a)–(f) *will be called a* coarse σ-field[9] (*on the* coarse sample space $(\Omega[i], \mathcal{F}[i], P[i])_{i=1}^{\infty}$). *Each set* A *belonging to the coarse σ-field* $\mathcal{A}$ *will be called* coarsely measurable (*w.r.t.* $\mathcal{A}$), *or a* coarse event.

Remark 2.4. Condition 2.3(c) is equivalent to

(c1) $\forall i \; \mathcal{F}[i] \subset \mathcal{A}$. That is, if a set $A \subset \Omega[\text{all}]$ is contained in some $\Omega[i]$, and is $\mathcal{F}[i]$-measurable, then $A \in \mathcal{A}$.

Also, Condition 2.3(d) is equivalent to each of the following conditions (d1)–(d4). There, we assume that $A \subset \Omega[\text{all}]$, $\forall i \; \left(A \cap \Omega[i]\right) \in \mathcal{F}[i]$, and $\forall k \; A_k \in \mathcal{A}$.

(d1) If $A_k \uparrow A$ (that is, $A_1 \subset A_2 \subset \ldots$ and $A = \cup_k A_k$) and $\sup_i P[i]\left((A \setminus A_k) \cap \Omega[i]\right) \to 0$ for $k \to \infty$, then $A \in \mathcal{A}$.

(d2) If $\sup_i P[i]\left((A \bigtriangleup A_k) \cap \Omega[i]\right) \to 0$ for $k \to \infty$, then $A \in \mathcal{A}$. (Here $A \bigtriangleup A_k = (A \setminus A_k) \cup (A_k \setminus A)$.)

(d3) If $A_k \uparrow A$ and $\limsup_i P[i]\left((A \setminus A_k) \cap \Omega[i]\right) \to 0$ for $k \to \infty$, then $A \in \mathcal{A}$.

(d4) If $\limsup_i P[i]\left((A \bigtriangleup A_k) \cap \Omega[i]\right) \to 0$ for $k \to \infty$, then $A \in \mathcal{A}$.

So, we have 10 equivalent combinations: (c)&(d), (c1)&(d), (c)&(d1), (c1)&(d1), (c)&(d2), ..., (c1)&(d4). (I omit the proof.)

However, "$\sup_i$" in (d) cannot be replaced with "$\limsup_i$".

---

[9] It is not a σ-field, unless $\mathcal{A}$ contains all sets satisfying 2.3(a).

Lemma 2.5. *Let* $\mathcal{A}_1$ *be a finite or countable set satisfying* 2.3(a,e) *and*

(b1) $\forall A, B \in \mathcal{A}_1 \ (A \cap B \in \mathcal{A}_1)$.

Then the least set $\mathcal{A}$ *containing* $\mathcal{A}_1$ *and satisfying* 2.3(b,c,d) *is a coarse* σ-*field.*

Proof. The algebra generated by $\mathcal{A}_1$ satisfies (e), since $P[i]\big((A \cup B) \cap \Omega[i]\big) = P[i](A \cap \Omega[i]) + P[i](B \cap \Omega[i]) - P[i]\big((A \cap B) \cap \Omega[i]\big)$. We enlarge the algebra according to (c), which preserves (e), as well as (a), (b). Finally, we enlarge it according to (d), which preserves (a), (b), (e); (c) and (f) hold trivially. □

In such a case we say that the coarse σ-field $\mathcal{A}$ is *generated* by the set $\mathcal{A}_1$.

Example 2.6. Let $\Omega[i] = \{0, \frac{1}{i}, \dots, \frac{i-1}{i}\}$, and $P[i]$ be the uniform distribution on $\Omega[i]$. Every interval $(s,t) \subset (0,1)$ gives us a set $A_{s,t} \subset \Omega[\text{all}]$,

$$A_{s,t} \cap \Omega[i] = (s,t) \cap \Omega[i].$$

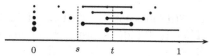

We take a dense countable set of pairs (s,t) (say, rational s,t) and consider the set $\mathcal{A}_1$ of the corresponding $A_{s,t}$. The set $\mathcal{A}_1$ satisfies the conditions of Lemma 2.5, therefore it generates a coarse σ-field $\mathcal{A}$. In fact, $\mathcal{A}$ consists of all $A = A[1] \uplus A[2] \uplus \dots$ such that sets $A[i] + (0, 1/i) \subset (0,1)$ converge in probability to some $A[\infty] \subset (0,1)$; that is, $\mathrm{mes}\big(A[\infty] \triangle (A[i] + (0,1/i))\big) \to 0$ for $i \to \infty$.

If $A = A_{s,t}$ then, of course, $A[\infty] = (s,t)$.

Example 2.7. Continuing Example 1.3, we take $\Omega[i] = \{-1, +1\}^i$ with the uniform distribution $P[i]$. Given n and $a = (a_1, \dots, a_n) \in \{-1, +1\}^n$, we consider $A_a \subset \Omega[\text{all}]$,

$$A_a \cap \Omega[i] = \{(\tau_1, \dots, \tau_i) : \tau_1 = a_1, \dots, \tau_n = a_n\} \quad \text{for } i \geq n.$$

Such sets A_a (for all a and n) are a countable collection $\mathcal{A}_1$ satisfying the conditions of Lemma 2.5, therefore it generates a coarse σ-field $\mathcal{A}$. In fact, $\mathcal{A}$ consists of all $A = A[1] \uplus A[2] \uplus \dots$ such that sets $\beta_i^{-1}(A) \subset (0,1)$ converge in probability to some $A[\infty] \subset (0,1)$; here $\beta_i : (0,1) \to \Omega[i]$ is such a measure preserving map:

$$\beta_i(x) = \big((-1)^{c_1}, \dots, (-1)^{c_i}\big) \quad \text{when } x - \left(\frac{c_1}{2} + \dots + \frac{c_i}{2^i}\right) \in \left(0, \frac{1}{2^i}\right),$$

for any $c_1, \dots, c_i \in \{0, 1\}$.

You may guess that some limiting procedure produces a ('true', not coarse) probability space out of any given coarse probability space. Indeed, such a procedure, called 'refinement', is described in Sect. 2.3.

2.3 Good Use of Joint Compactification

Having a coarse probability space $((\Omega[i], \mathcal{F}[i], P[i])_{i=1}^{\infty}, \mathcal{A})$ and its refinement $(\Omega, \mathcal{F}, P)$ (to be defined later), we may hope that the Hilbert space $L_2[\infty] = L_2(\Omega, \mathcal{F}, P)$ is in some sense the limit of Hilbert spaces $L_2[i] = L_2(\Omega[i], \mathcal{F}[i], P[i])$. That is indeed the case in the framework of joint compactification, as we'll see. A bad use of the framework, tried in Sect. 1.2, is a joint compactification of given probability spaces. A good use, considered here, is a joint compactification of metric (Hilbert, ...) spaces built over the given probability spaces.

Definition 2.8. *A coarse Polish space is* $((S[i], \rho[i])_{i=1}^{\infty}, c)$, *where each* $(S[i], \rho[i])$ *is a Polish space (that is, a complete separable metric space[10]), and* $c \subset S[1] \times S[2] \times \dots$ *is a set of sequences* $x = (x[1], x[2], \dots)$ *satisfying the following conditions:*

(a) *if* $x_1, x_2 \in S[1] \times S[2] \times \dots$ *are such that* $\rho[i](x_1[i], x_2[i]) \to 0$ *(for* $i \to \infty$*), then* $(x_1 \in c) \Longleftrightarrow (x_2 \in c)$;

(b) *if* $x, x_1, x_2, \dots \in S[1] \times S[2] \times \dots$ *are such that* $\sup_i \rho[i](x_k[i], x[i]) \to 0$ *(for* $k \to \infty$*), then* $(\forall k \; x_k \in c) \Longrightarrow (x \in c)$;

(c) $\lim_i \rho[i](x_1[i], x_2[i])$ *exists for every* $x_1, x_2 \in c$;

(d) *there exists a finite or countable subset* $c_1 \subset c$ *that generates* c *in the sense that the least subset of* c *satisfying* (a), (b) *and containing* c_1 *is the whole* c.

Remark 2.9. Condition 2.8(d) does not change if 'satisfying (a), (b)' is replaced with 'satisfying (b)'. That is, 2.8(d) is just separability of c in the metric $x_1, x_2 \mapsto \sup_i \rho[i](x_1[i], x_2[i])$.

The refinement of a coarse Polish space $((S[i], \rho[i])_{i=1}^{\infty}, c)$ is basically the metric space $(c, \tilde{\rho})$, where

$$\tilde{\rho}(x_1, x_2) = \lim_i \rho[i](x_1[i], x_2[i]).$$

However, $\tilde{\rho}$ is a pseudometric (semimetric); it may vanish for some $x_1 \neq x_2$. The equivalence class, denoted by $x[\infty]$, of a sequence $x \in c$ consists of all $x_1 \in c$ such that $\rho[i](x_1[i], x[i]) \to 0$. On the set $S[\infty]$ of all equivalence classes we introduce a metric $\rho[\infty]$,

$$\rho[\infty](x_1[\infty], x_2[\infty]) = \lim_{i \to \infty} \rho[i](x_1[i], x_2[i]);$$

thus, $(S[\infty], \rho[\infty])$ is a metric space. We write

$$(S[\infty], \rho[\infty]) = \mathrm{Lim}_{i \to \infty, c}(S[i], \rho[i])$$

---

[10] Many authors define a Polish space as a metrizable topological space admitting a complete separable metric. However, I assume that a metric is given.

and call $(S[\infty], \rho[\infty])$ the *refinement* of the coarse Polish space $((S[i], \rho[i])_{i=1}^{\infty}, c)$. Also, for every $x = (x[1], x[2], \dots) \in c$ we denote its equivalence class $x[\infty] \in S[\infty]$ by

$$x[\infty] = \mathrm{Lim}_{i \to \infty, c} \, x[i] \,,$$

and call it the refinement of x.

Lemma 2.10. *For every coarse Polish space, its refinement (S, ρ) is a Polish space.*

Proof. Separability follows from 2.8(d); completeness is to be proven. Let $x_1, x_2, \dots$ be a Cauchy sequence in (S, ρ); we have to find $x \in S$ such that $\rho(x_k, x) \to 0$. We may assume that $\sum_k \rho(x_k, x_{k+1}) < \infty$. Each x_k is an equivalence class; using (a) we choose for each $k = 1, 2, 3, \dots$ a representative $s_k \in S[1] \times S[2] \times \dots$ of x_k such that $\sup_i \rho[i] \big(s_k[i], s_{k+1}[i] \big) \le 2\rho(x_k, x_{k+1})$. Completeness of $(S[i], \rho[i])$ ensures existence of $s_\infty[i] = \lim_k s_k[i]$. Condition (b) ensures $s_\infty \in c$. The equivalence class $x \in S$ of s_∞ satisfies $\rho(x_k, x) \le \sup_i \rho[i] \big(s_k[i], s_\infty[i] \big) \to 0$ for $k \to \infty$. $\qquad\square$

Let $(S[i], \rho[i])_{i=1}^{\infty}, c)$ be a coarse Polish space, and (S, ρ) its refinement. On the disjoint union $(S[1] \uplus S[2] \uplus \dots) \uplus S$ we introduce a topology, namely, the weakest topology making continuous the following functions $f_s : (S[1] \uplus S[2] \uplus \dots) \uplus S \to [0, \infty)$ for $s \in c$,

$$f_s(x) = \rho[i] \big(x, s[i] \big) \quad \text{for } x \in S[i] \,,$$
$$f_s(x) = \rho(x, s[\infty]) \quad \text{for } x \in S \,,$$

and an additional function $f_0 : (S[1] \uplus S[2] \uplus \dots) \uplus S \to [0, \infty)$, $f_0(x) = 1/i$ for $x \in S[i]$, $f_0(x) = 0$ for $x \in S$. On every $S[i]$ separately (and also on S), the new topology coincides with the old topology, given by $\rho[i]$ (or ρ).

We may choose a sequence (s_k) dense in c; the topology is generated by functions f_{s_k} (and f_0), therefore it is a metrizable topology. Moreover, the sequence of functions $\big(\frac{f_{s_k}(\cdot)}{1 + f_{s_k}(\cdot)} \big)_{k=1}^{\infty}$ (and f_0) maps the disjoint union into the metrizable compact space $[0, 1]^{\infty}$, and is a homeomorphic embedding. Thus, we have a joint compactification of all $S[i]$ and S; and so, we treat them as subsets of a compact metrizable space K;

$$S[i] \subset K, \quad S \subset K \,.$$

Lemma 2.11. *Let $s_\infty \in S$, $s_1 \in S[1]$, $s_2 \in S[2], \dots$ Then $s_i \to s_\infty$ in K if and only if $s = (s_1, s_2, \dots) \in c$ and $\mathrm{Lim}_{i \to \infty, c} \, s_i = s_\infty$.*

Proof. The 'if' part. The needed relation, $f_k(s_i) \to f_k(s_\infty)$ for $i \to \infty$, is ensured by 2.8(c).

The 'only if' part. We choose $x \in c$ such that $x[\infty] = s_\infty$; then $\rho[i] \big(s_i, x[i] \big) \to \rho \big(s_\infty, x[\infty] \big) = 0$, thus $s \in c$ by 2.8(a). $\qquad\square$

The assumption '$s_\infty \in S$' is essential. Other limiting points (not belonging to S) may exist; corresponding sequences converge in K but do not belong to c. And, of course, sets $S, S[1], S[2], \ldots$ are not closed in K, unless they are compact.

Lemma 2.12. *A set $c_1 \subset c$ generates c if and only if the set of refinements $\{x[\infty] : x \in c_1\}$ is dense in $S[\infty]$.*

Proof. The 'only if' part follows from a simple argument: if S' is a closed subset of S then the set c' of all $x \in c$ such that $x[\infty] \in S'$ satisfies 2.8(a,b).

The 'if' part. Let $\{x[\infty] : x \in c_1\}$ be dense in $S[\infty]$ and $s \in c$. We choose $x_k \in c_1$ such that $x_k[\infty] \to s$. Similarly to the proof of Lemma 2.10, we construct $y_k \in c_1$ such that $\rho[i]\big(s_k[i], y_k[i]\big) \to 0$ when $i \to \infty$ for each k, and $\sup_i \rho[i]\big(y_k[i], s[i]\big) \to 0$ when $k \to \infty$. The subset of c generated by c_1 contains all y_k by 2.8(a). Thus, it contains s by 2.8(b). $\square$

Given continuous functions $f[i] : S[i] \to \mathbb{R}$, $f[\infty] : S[\infty] \to \mathbb{R}$, we write $f[\infty] = \mathrm{Lim}_{i\to\infty,c} f[i]$ if $f[i](x[i]) \to f[\infty](x[\infty])$ whenever $x[\infty] = \mathrm{Lim}_{i\to\infty,c} x[i]$. If functions $f[i]$ are equicontinuous (say, $|f[i](x) - f[i](y)| \le \rho[i](x,y)$ for all i and $x, y \in S[i]$), then it is enough to check that $f[i](x_k[i]) \to f[\infty](x_k[\infty])$ for some sequence $(x_k)_{k=1}^\infty$, $x_k \in c$, such that the sequence $(x_k[\infty])_{k=1}^\infty$ is dense in $S[\infty]$.

Given continuous maps $f[i] : S[i] \to S[i]$, $f[\infty] : S \to S$, we write $f[\infty] = \mathrm{Lim}_{i\to\infty,c} f[i]$ if $\mathrm{Lim}_{i\to\infty,c} f[i](x[i]) = f[\infty](x[\infty])$ whenever $x[\infty] = \mathrm{Lim}_{i\to\infty,c} x[i]$. That is, $\mathrm{Lim}\big(f[i](x[i])\big) = \big(\mathrm{Lim}\, f[i]\big)\big(\mathrm{Lim}\, x[i]\big)$. If maps $f[i]$ are equicontinuous then, again, convergence may be checked on x_k such that $x_k[\infty]$ are dense.

Given continuous maps $f[i] : S[\infty] \to S[i]$, we may ask whether $\mathrm{Lim}_{i\to\infty,c} f[i](x) = x$ for all $x \in S[\infty]$, or not. If maps $f[i]$ are equicontinuous then, still, convergence may be checked for a dense subset of $S[\infty]$.

If every $S[i]$ is not only a metric space but also a Hilbert (or Banach) space, and c is linear (that is, closed under linear operations), then the refinement S is also a Hilbert (or Banach) space, and linear operations are continuous on $\big(S[1] \cup S[2] \cup \ldots\big) \cup S \subset K$ in the sense that

$$\mathrm{Lim}_{i\to\infty,c}(as_1[i] + bs_2[i]) = a\,\mathrm{Lim}_{i\to\infty,c}\, s_1[i] + b\,\mathrm{Lim}_{i\to\infty,c}\, s_2[i]$$

for all $s_1, s_2 \in c$.

Consider the case of Hilbert spaces $S[i] = H[i]$, $S = H$. Given linear[11] operators $R[i] : H[i] \to H[i]$, we may ask about $\mathrm{Lim}\, R[i]$. If it exists, we get

$$\mathrm{Lim}\big(R[i]x[i]\big) = \big(\mathrm{Lim}\, R[i]\big)\big(\mathrm{Lim}\, x[i]\big).$$

If $\sup_i \|R[i]\| < \infty$, then $R[i]$ are equicontinuous, and convergence may be checked on a sequence x_k such that vectors $x_k[\infty]$ span H (that is, their

---

[11] Continuous, of course.

linear combinations are dense in H). For example, one-dimensional orthogonal projections; if $x[\infty] = \operatorname{Lim} x[i]$ then $\operatorname{Proj}_{x[\infty]} = \operatorname{Lim} \operatorname{Proj}_{x[i]}$.

Given linear operators $R[i] : H \to H[i]$, we may ask whether $\operatorname{Lim} R[i](x) = x$ for all $x \in H$, or not. If $\sup_i \|R[i]\| < \infty$ then convergence may be checked on a sequence that spans H. Such $R[i]$ always exist; moreover, $\|R[i]\| \leq 1$ may be ensured. Proof: we take x_k such that $x_k[\infty]$ are an orthonormal basis of H. After some correction, $x_k[i]$ become orthogonal (for each i), and $\|x_k(i)\| \leq 1$.[12] Now we let $R[i] x_k[\infty] = x_k[i]$.

We return to coarse *probability* spaces.

Let $\left((\Omega[i], \mathcal{F}[i], P[i])_{i=1}^{\infty}, \mathcal{A} \right)$ be a coarse probability space. For each i the pseudometric $A, B \mapsto P[i](A \triangle B)$ on $\mathcal{F}[i]$ gives us the metric space $\operatorname{MALG}[i] = \operatorname{MALG}\big(\Omega[i], \mathcal{F}[i], P[i]\big)$ of all equivalence classes of measurable sets. It is not only a metric space but also a Boolean algebra, and moreover, a separable measure algebra (as defined in [7, 17.44]). Treating every coarse event $A \in \mathcal{A}$ as a sequence of $A[1] \in \operatorname{MALG}[1], A[2] \in \operatorname{MALG}[2], \ldots$ we get a coarse Polish space $((\operatorname{MALG}[i])_{i=1}^{\infty}, \mathcal{A})$. Its refinement is a metric space $\operatorname{MALG}[\infty]$. The set $\mathcal{A}$ is closed under Boolean operations (union, intersection, complement). Therefore $\operatorname{MALG}[\infty]$ is not only a metric space but also a Boolean algebra. Using Lemma 2.10 it is easy to check that $\operatorname{MALG}[\infty]$ is a separable measure algebra. Therefore [7, 17.44] it is (up to isomorphism) of the form

$$\operatorname{MALG}[\infty] = \operatorname{MALG}(\Omega, \mathcal{F}, P)$$

for some probability space $(\Omega, \mathcal{F}, P)$. In the nonatomic case we may take $(\Omega, \mathcal{F}, P) = (0, 1)$ with Lebesgue measure; in general, we may take a shorter (maybe, empty) interval plus a finite (maybe, empty) or countable set of atoms. Such a probability space $(\Omega.\mathcal{F}, P)$ (unique up to isomorphism) will be called the refinement of the coarse probability space $\left((\Omega[i], \mathcal{F}[i], P[i])_{i=1}^{\infty}, \mathcal{A} \right)$, and we write

$$(\Omega, \mathcal{F}, P) = \operatorname{Lim}_{i \to \infty, \mathcal{A}} \big(\Omega[i], \mathcal{F}[i], P[i] \big)$$

(in practice, sometimes I omit "$i \to \infty$" or "$\mathcal{A}$" or both under the "Lim").

Every sequence $A = (A[1], A[2], \ldots) \in \mathcal{A}$ has its refinement

$$\operatorname{Lim}_{i \to \infty, \mathcal{A}} A[i] = A[\infty] \in \operatorname{MALG}(\Omega, \mathcal{F}, P).$$

Lemma 2.13. *A subset $\mathcal{A}_1$ of a coarse σ-field $\mathcal{A}$ generates $\mathcal{A}$ if and only if the refinement $\mathcal{F}$ of $\mathcal{A}$ is generated* (mod 0) *by refinements $A[\infty]$ of all $A \in \mathcal{A}_1$.*

Proof. We apply Lemma 2.12 to the algebra generated by $\mathcal{A}_1$. $\qquad\square$

In order to define $L_2(\mathcal{A})$ as a set of functions on $\Omega[\text{all}]$, we start with indicators $\mathbf{1}_A$ for $A \in \mathcal{A}$, form their linear combinations, and take their completion in the metric

---

[12] Of course, $\|x_k[i]\| \to 1$ for $i \to \infty$, but in general we cannot ensure $\|x_k[i]\| = 1$. It may happen that $\dim H[i] < \infty$ but $\dim H = \infty$.

$$\|f\|_{L_2(\mathcal{A})} = \sup_i \|f[i]\|_{L_2[i]},$$

where $L_2[i] = L_2(\Omega[i], \mathcal{F}[i], P[i])$; the completion is a Banach (not Hilbert) space $L_2(\mathcal{A})$. Each element f of the completion is evidently identified with a sequence of $f[i] \in L_2[i]$, or a function on $\Omega[\text{all}]$. We have a coarse Polish space $((L_2[i])_{i=1}^{\infty}, L_2(\mathcal{A}))$. It has its refinement, $L_2[\infty]$.

Lemma 2.14. *The refinement $L_2[\infty]$ of $((L_2[i])_{i=1}^{\infty}, L_2(\mathcal{A}))$ is (canonically isomorphic to) $L_2(\Omega, \mathcal{F}, P)$, where $(\Omega, \mathcal{F}, P)$ is the refinement of $((\Omega[i], \mathcal{F}[i], P[i])_{i=1}^{\infty}, \mathcal{A})$.*

Proof. We define the canonical map $L_2(\mathcal{A}) \to L_2(\Omega, \mathcal{F}, P)$ first on indicators by $1_A \mapsto 1_{A[\infty]}$, and extend it by linearity and continuity to the whole $L_2(\mathcal{A})$. We note that the image of $f \in L_2(\mathcal{A})$ in $L_2(\Omega, \mathcal{F}, P)$ depends only on the refinement $f[\infty] \in L_2[\infty]$ of f, and their norms are equal (both are equal to $\lim_i \|f[i]\|$). We have a linear isometric embedding $L_2[\infty] \to L_2(\Omega, \mathcal{F}, P)$. Its image is closed (since $L_2[\infty]$ is complete by Lemma 2.10), and contains indicators 1_B for all $B \in \text{MALG}(\Omega, \mathcal{F}, P)$; therefore the image is the whole $L_2(\Omega, \mathcal{F}, P)$. $\square$

Remark 2.15. The same holds for L_p for each $p \in (0, \infty)$, and for the space L_0 of all random variables (equipped with the topology of convergence in probability). Elements of $L_0(\mathcal{A})$ will be called *coarsely measurable* (w.r.t. $\mathcal{A}$) functions (on $\Omega[\text{all}]$), or *coarse random variables;* elements of $L_2(\mathcal{A})$ — square integrable coarse random variables.

Let f be a coarse random variable. Then (usual) random variables $f[i] : \Omega[i] \to \mathbb{R}$ converge in distribution (for $i \to \infty$) to the refinement $f[\infty] : \Omega \to \mathbb{R}$. The distribution of $f[\infty]$ will be called the *limiting distribution* of f.

It may happen that $f \in L_2(\mathcal{A})$ but $(\text{sgn} f) \notin L_2(\mathcal{A})$. An example: $f(\omega) = \frac{(-1)^i}{i}$ for all $\omega \in \Omega[i]$. Here, the limiting distribution is an atom at 0, and the function 'sgn' is discontinuous at 0.

Lemma 2.16. (a) *Let $f : \Omega[\text{all}] \to \mathbb{R}$ be a coarse random variable, and $\varphi : \mathbb{R} \to \mathbb{R}$ a continuous function. Then $\varphi \circ f : \Omega[\text{all}] \to \mathbb{R}$ is a coarse random variable.*

(b) *The same as (a) but φ may be discontinuous at points of a set $Z \subset \mathbb{R}$, negligible w.r.t. the limiting distribution of f.*

Proof. If f is a linear combination of indicators, then $\varphi \circ f$ is another linear combination of the same indicators. A straightforward approximation gives (a) for uniformly continuous φ. In general, for every ε there exists a compact set $K \subset \mathbb{R} \setminus Z$ of probability $\geq 1 - \varepsilon$ w.r.t. the limiting distribution, and also w.r.t. the distribution of $f[i]$ for all i (since all these distributions are a compact set of distributions). The restriction of f to K is uniformly continuous. The limit for $\varepsilon \to 0$ is uniform in i. $\square$

For a given Polish space S we may define a coarse S-valued random variable as a map $f : \Omega[\text{all}] \to S$ such that (usual) random variables $f[i] : \Omega[i] \to S$ converge in distribution (for $i \to \infty$), and $f^{-1}(B) \in \mathcal{A}$ for every $B \subset S$ such that the boundary of B is negligible w.r.t. the limiting distribution of f.

For $S = \mathbb{R}$ the new definition conforms with the old one.

A coarse σ-field generated by a given sequence of sets (coarse events) was defined after Lemma 2.5. Often it is convenient to generate a coarse σ-field by a sequence of functions (coarse random variables). A function $f : \Omega[\text{all}] \to \mathbb{R}$ is coarsely $\mathcal{A}$-measurable if and only if $\mathcal{A}$ contains sets $f^{-1}((-\infty, x))$ for all $x \in \mathbb{R}$ except for atoms (if any) of the limiting distribution of f. A dense countable subset of these x is enough. So, a coarse σ-field generated by a finite or countable set of functions f is nothing but the coarse σ-field generated by a countable set of sets of the form $f^{-1}((-\infty, x))$. More generally, S-valued (coarse) random variables may be used; they are reduced to the real-valued case by composing with appropriate continuous functions $S \to \mathbb{R}$.

Lemma 2.17. *A sequence of functions $f_k : \Omega[\text{all}] \to \mathbb{R}$ generates a coarse σ-field if and only if for every n, n-dimensional random variables $\big(f_1[i], \ldots, f_n[i]\big) : \Omega[i] \to \mathbb{R}^n$ converge in distribution (for $i \to \infty$).*

Proof. The 'only if' part. Let $f_1, \ldots, f_n$ be coarsely measurable (w.r.t. some coarse σ-field), then they have a limiting joint distribution.

The 'if' part. For each n we choose a dense countable set $Q_n \subset \mathbb{R}$ negligible w.r.t. the limiting distribution of f_n. We apply Lemma 2.5 to the set $\mathcal{A}_1$ of coarse events of the form $\{f_1(\cdot) \leq q_1, \ldots, f_n(\cdot) \leq q_n\}$ where $q_1 \in Q_1, \ldots, q_n \in Q_n$, $n = 1, 2, \ldots$ □

Remark 2.18. The same holds for an arbitrary Polish space instead of $\mathbb{R}$.

Remark 2.19. Comparing Lemma 2.17 and (1.7) we see that every joint compactification of $\Omega_1 \uplus \Omega_2 \uplus \ldots$ (in the sense of Sect. 1.2, assuming (1.6)) may be downgraded to a coarse probability space. Namely, we take a sequence of functions f_k that generates C and consider the coarse σ-field $\mathcal{A}$ generated by (f_k). Every $f \in C$ is a coarse random variable, since $L_0(\mathcal{A})$ is closed under all operations used in (1.3), (1.4), or (1.5).[13] Therefore $\mathcal{A}$ does not depend on the choice of (f_k).

---

[13] Of course, $L_0(\mathcal{A})$ usually contains no sequence dense in the *uniform* topology.

3 Scaling Limit and Independence

3.1 Product of Coarse Probability Spaces

Having two coarse probability spaces $((\Omega_1[i], \mathcal{F}_1[i], P_1[i])_{i=1}^{\infty}, \mathcal{A}_1)$ and $((\Omega_2[i], \mathcal{F}_2[i], P_2[i])_{i=1}^{\infty}, \mathcal{A}_2)$, we define their product as the coarse probability space $((\Omega[i], \mathcal{F}[i], P[i])_{i=1}^{\infty}, \mathcal{A})$ where for each i,

$$(\Omega[i], \mathcal{F}[i], P[i]) = (\Omega_1[i], \mathcal{F}_1[i], P_1[i]) \times (\Omega_2[i], \mathcal{F}_2[i], P_2[i])$$

is the usual product of probability spaces, and $\mathcal{A}$ is the smallest coarse σ-field that contains $\{A_1 \times A_2 : A_1 \in \mathcal{A}_1, A_2 \in \mathcal{A}_2\}$, where $A_1 \times A_2 \subset \Omega[\text{all}]$ is defined by $\forall i \ (A_1 \times A_2)[i] = A_1[i] \times A_2[i]$. Existence of such $\mathcal{A}$ is ensured by Lemma 2.5. We write $\mathcal{A} = \mathcal{A}_1 \otimes \mathcal{A}_2$.

Lemma 3.1. *The refinement of the product of two coarse probability spaces is (canonically isomorphic to) the product of their refinements.*

Proof. Denote these refinements by $(\Omega_1, \mathcal{F}_1, P_1)$, $(\Omega_2, \mathcal{F}_2, P_2)$ and $(\Omega, \mathcal{F}, P)$. Both $\text{MALG}(\Omega_1, \mathcal{F}_1, P_1)$ and $\text{MALG}(\Omega_2, \mathcal{F}_2, P_2)$ are naturally embedded into $\text{MALG}(\Omega, \mathcal{F}, P)$ as *independent* subalgebras. They generate $\text{MALG}(\Omega, \mathcal{F}, P)$ due to Lemma 2.13. $\qquad\square$

Given an arbitrary coarse σ-field $\mathcal{A}$ on the product coarse sample space $((\Omega_1[i], \mathcal{F}_1[i], P_1[i]) \times (\Omega_2[i], \mathcal{F}_2[i], P_2[i]))_{i=1}^{\infty}$, we may ask whether $\mathcal{A}$ is a product, that is, $\mathcal{A} = \mathcal{A}_1 \otimes \mathcal{A}_2$ for some $\mathcal{A}_1, \mathcal{A}_2$, or not. No need to check all $\mathcal{A}_1, \mathcal{A}_2$. Rather, we have to check

$$\mathcal{A}_1 = \{A_1 : A_1 \times \Omega_2 \in \mathcal{A}\}, \quad \mathcal{A}_2 = \{A_2 : \Omega_1 \times A_2 \in \mathcal{A}\};$$

of course, $A_1 \times \Omega_2 \subset \Omega[\text{all}]$ is defined by $\forall i \ (A_1 \times \Omega_2)[i] = A_1[i] \times \Omega_2[i]$. If $\{A_1 \times A_2 : A_1 \in \mathcal{A}_1, A_2 \in \mathcal{A}_2\}$ generates $\mathcal{A}$, then $\mathcal{A}$ is a product; otherwise, it is not.

The refinement $\mathcal{F}$ of $\mathcal{A}$ contains two sub-σ-fields $\mathcal{F}_1 = \{(A_1 \times \Omega_2)[\infty] : A_1 \in \mathcal{A}_1\}$, $\mathcal{F}_2 = \{(\Omega_1 \times A_2)[\infty] : A_2 \in \mathcal{A}_2\}$. They are independent:

$$P(A \cap B) = P(A)P(B) \quad \text{for } A \in \mathcal{F}_1, B \in \mathcal{F}_2.$$

Lemma 3.2. *$\mathcal{A}$ is a product if and only if $\mathcal{F}_1, \mathcal{F}_2$ generate $\mathcal{F}$.*

Proof. We apply Lemma 2.13 to $\{A_1 \times A_2 : A_1 \in \mathcal{A}_1, A_2 \in \mathcal{A}_2\}$. $\qquad\square$

Remark 3.3. It is well-known that a generating pair of independent sub-σ-fields means that $(\Omega, \mathcal{F}, P)$ is (isomorphic to) the product of two probability spaces. So, a coarse probability space is a product if and only if its refinement is a product. (Assuming, of course, that the coarse *sample* space is a product.)

Let $\mathcal{A} = \mathcal{A}_1 \otimes \mathcal{A}_2$. Consider Hilbert spaces $H_1[i] = L_2(\Omega_1[i], \mathcal{F}_1[i], P_1[i])$, $H_2[i] = L_2(\Omega_2[i], \mathcal{F}_2[i], P_2[i])$, $H[i] = L_2(\Omega[i], \mathcal{F}[i], P[i])$. For each i, the space $H[i]$ is (canonically isomorphic to) $H_1[i] \otimes H_2[i]$. Indeed, for $x_1 \in H_1[i]$, $x_2 \in H_2[i]$ we define $x_1 \otimes x_2 \in H[i]$ by $(x_1 \otimes x_2)(\omega_1, \omega_2) = x_1(\omega_1) x_2(\omega_2)$; then $\langle x_1 \otimes x_2, y_1 \otimes y_2 \rangle = \langle x_1, y_1 \rangle \langle x_2, y_2 \rangle$, and factorizable vectors (of the form $x_1 \otimes x_2$) span the space $H[i]$. We know (see Lemma 2.14) that the refinement $H[\infty]$ of $\big((H[i])_{i=1}^{\infty}, L_2(\mathcal{A})\big)$ is $L_2(\Omega, \mathcal{F}, P)$. Also, $H_1[\infty] = L_2(\Omega_1, \mathcal{F}_1, P_1)$ and $H_2[\infty] = L_2(\Omega_2, \mathcal{F}_2, P_2)$. Using Lemma 3.1 we get $H[\infty] = H_1[\infty] \otimes H_2[\infty]$. In that sense,

$$\mathrm{Lim}\big(H_1[i] \otimes H_2[i]\big) = \big(\mathrm{Lim}\, H_1[i]\big) \otimes \big(\mathrm{Lim}\, H_2[i]\big).$$

If $x \in L_2(\mathcal{A}_1)$, $y \in L_2(\mathcal{A}_2)$, we define $x \otimes y$ by $(x \otimes y)[i] = x[i] \otimes y[i]$ for all i. We get $x \otimes y \in L_2(\mathcal{A})$ and $(x \otimes y)[\infty] = x[\infty] \otimes y[\infty]$, that is,

$$\mathrm{Lim}\big(x[i] \otimes y[i]\big) = \big(\mathrm{Lim}\, x[i]\big) \otimes \big(\mathrm{Lim}\, y[i]\big), \tag{3.1}$$

since it holds for (linear combinations of) indicators of coarse events. Note also that linear combinations of factorizable vectors are dense in $L_2(\mathcal{A})$.

Assume that $R_1[i] : H_1[i] \to H_1[i]$, $R_2[i] : H_2[i] \to H_2[i]$ are linear operators, possessing limits $R_1[\infty] = \mathrm{Lim}\, R_1[i]$, $R_2[\infty] = \mathrm{Lim}\, R_2[i]$. Consider linear operators $R_1[i] \otimes R_2[i] = R[i] : H[i] \to H[i]$. (It means that $R[i]x[i] = R_1[i]x_1[i] \otimes R_2[i]x_2[i]$ whenever $x[i] = x_1[i] \otimes x_2[i]$.) If $\sup_i \|R_1[i]\| < \infty$, $\sup_i \|R_2[i]\| < \infty$, then $\mathrm{Lim}\, R[i] = R_1[\infty] \otimes R_2[\infty]$, that is,

$$\mathrm{Lim}\big(R_1[i] \otimes R_2[i]\big) = \big(\mathrm{Lim}\, R_1[i]\big) \otimes \big(\mathrm{Lim}\, R_2[i]\big). \tag{3.2}$$

Proof: We have to check that

$$\mathrm{Lim}\big(R_1[i] \otimes R_2[i]\big)x[i] = \big(\mathrm{Lim}\, R_1[i] \otimes \mathrm{Lim}\, R_2[i]\big)\big(\mathrm{Lim}\, x[i]\big)$$

for all $x \in L_2(\mathcal{A})$. We may assume that x is factorizable, $x = x_1 \otimes x_2$; then

$$\mathrm{Lim}\big(R_1[i] \otimes R_2[i]\big)\big(x_1[i] \otimes x_2[i]\big) =$$
$$= \mathrm{Lim}\big(R_1[i]x_1[i] \otimes R_2[i]x_2[i]\big) =$$
$$= \big(\mathrm{Lim}\, R_1[i]x_1[i]\big) \otimes \big(\mathrm{Lim}\, R_2[i]x_2[i]\big) =$$
$$= \big(\mathrm{Lim}\, R_1[i]\big)\big(\mathrm{Lim}\, x_1[i]\big) \otimes \big(\mathrm{Lim}\, R_2[i]\big)\big(\mathrm{Lim}\, x_2[i]\big) =$$
$$= \big(\mathrm{Lim}\, R_1[i] \otimes \mathrm{Lim}\, R_2[i]\big)\big(\mathrm{Lim}\, x_1[i] \otimes \mathrm{Lim}\, x_2[i]\big).$$

Especially, let $R_2[i]$ be the orthogonal projection to the one-dimensional subspace of constants (basically, the expectation), and $R_1[i]$ be the unit (identity) operator. Then $\big(R_1[i] \otimes R_2[i]\big)\big(x[i]\big) = \mathbb{E}\big(x[i] \,\big|\, \mathcal{F}_1[i]\big)$, since it holds for factorizable vectors. Further, $R_2[\infty] = \mathrm{Lim}\, R_2[i]$ is the expectation on $(\Omega_2, \mathcal{F}_2, P_2)$, since convergence of vectors implies convergence of one-dimensional projections, and constant functions on $\Omega_2[\mathrm{all}]$ belong to $L_2(\mathcal{A})$. So,

$$\text{Lim}\,\mathbb{E}\left(\,x[i]\,|\,\mathcal{F}_1[i]\,\right) = \mathbb{E}\left(\,\text{Lim}\,x[i]\,|\,\mathcal{F}_1\,\right) \qquad (3.3)$$

for all $x \in L_2(\mathcal{A})$.

All the same holds for the product of any finite number of spaces (not just two).

3.2 Dyadic Case

Let $(\Omega[i], \mathcal{F}[i], P[i])$ be the space of all maps $\frac{1}{i}\mathbb{Z} \to \{-1, +1\}$ with the usual product measure. That is, we have independent random signs $\tau_{k/i}$ for all integers k;[14] each random sign takes on two values ± 1 with probabilities $50\%, 50\%$. The coarse sample space $(\Omega[i], \mathcal{F}[i], P[i])_{i=1}^{\infty}$ will be called the *dyadic coarse sample space*.[15] Let $\mathcal{A}$ be a coarse σ-field on the dyadic coarse sample space. What about decomposing it, say, into the past and the future w.r.t. a given instant?

Let us define a *coarse instant* as a sequence $t = (t[i])_{i=1}^{\infty}$ such that $t[i] \in \frac{1}{i}\mathbb{Z}$ (that is, $it[i] \in \mathbb{Z}$) for all i, and there exists $t[\infty] \in \mathbb{R}$ (call it the refinement of the coarse instant) such that $t[i] \to t[\infty]$ for $i \to \infty$. A *coarse time interval* is a pair (s, t) of coarse instants s, t such that $s \le t$ in the sense that $s[i] \le t[i]$ for all i.

For every coarse time interval (s, t) we define the coarse probability space $((\Omega_{s,t}[i], \mathcal{F}_{s,t}[i], P_{s,t}[i])_{i=1}^{\infty}, \mathcal{A}_{s,t})$ as follows. First, $\Omega_{s,t}[i]$ is the space of all maps $\left(\frac{1}{i}\mathbb{Z} \cap [s[i], t[i]]\right) \to \{-1, +1\}$.[16] Second, $\mathcal{F}_{s,t}[i]$ and $P_{s,t}[i]$ are defined naturally, and we have the canonical measure preserving map $(\Omega[i], \mathcal{F}[i], P[i]) \to (\Omega_{s,t}[i], \mathcal{F}_{s,t}[i], P_{s,t}[i])$. Third, each $A \subset \Omega_{s,t}[\text{all}]$ has its inverse image in $\Omega[\text{all}]$; if the inverse image of A belongs to $\mathcal{A}$ then (and only then) A belongs to $\mathcal{A}_{s,t}$, which is the definition of $\mathcal{A}_{s,t}$. It is easy to see that $\mathcal{A}_{s,t}$ is a coarse σ-field.

Given coarse time intervals (r, s) and (s, t), we have

$$\left(\Omega_{r,t}[i], \mathcal{F}_{r,t}[i], P_{r,t}[i]\right) = \left(\Omega_{r,s}[i], \mathcal{F}_{r,s}[i], P_{r,s}[i]\right) \times \left(\Omega_{s,t}[i], \mathcal{F}_{s,t}[i], P_{s,t}[i]\right),$$

and we may ask whether $\mathcal{A}_{r,t}$ is a product, that is, $\mathcal{A}_{r,t} = \mathcal{A}_{r,s} \otimes \mathcal{A}_{s,t}$, or not.

Definition 3.4. *A dyadic coarse factorization is a coarse probability space* $((\Omega[i], \mathcal{F}[i], P[i])_{i=1}^{\infty}, \mathcal{A})$ *such that* $(\Omega[i], \mathcal{F}[i], P[i])_{i=1}^{\infty}$ *is the dyadic coarse sample space;*

$$\mathcal{A}_{r,t} = \mathcal{A}_{r,s} \otimes \mathcal{A}_{s,t}$$

whenever r, s, t *are coarse instants such that* $r[i] \le s[i] \le t[i]$ *for all* i*; and*

---

[14] Rigorously, I should denote it by $\tau_k[i]$, but $\tau_{k/i}$ is more expressive. Though $\tau_{2/6}$ is not the same as $\tau_{1/3}$, hopefully, it does not harm.

[15] Sometimes a subsequence is used; say, $i \in \{2, 4, 8, 16, \dots\}$ only; or equivalently, $\Omega[i]$ is the space of maps $2^{-i}\mathbb{Z} \to \{-1, +1\}$; see Examples 3.9, 3.10.

[16] It may happen that $s[i] = t[i]$, then $\Omega_{s,t}[i]$ contains a single point.

$$\mathcal{A} \text{ is generated by } \bigcup_{(s,t)} \mathcal{A}_{s,t},$$

where the union is taken over all coarse time intervals (s,t).

Example 3.5. A single function $f : \Omega[\text{all}] \to \mathbb{R}$, defined by $f(\omega) = \tau_{0/i}(\omega)$ for $\omega \in \Omega[i]$, generates a coarse σ-field $\mathcal{A}$. However, the coarse probability space $\left((\Omega[i], \mathcal{F}[i], P[i])_{i=1}^{\infty}, \mathcal{A}\right)$ is *not* a dyadic coarse factorization. The equality $\mathcal{A}_{r,t} = \mathcal{A}_{r,s} \otimes \mathcal{A}_{s,t}$ is violated when $s[i]$ converges to 0 from both sides; say, $s[i] = (-1)^i/i$. It means that a single point of the time continuum should not carry a random sign. See also Lemmas 3.11–3.13.

Every family $(\mathcal{A}_{s,t})_{s \leq t}$ of coarse σ-fields $\mathcal{A}_{s,t}$ on coarse sample spaces $\left(\Omega_{s,t}[i], \mathcal{F}_{s,t}[i], P_{s,t}[i]\right)_{i=1}^{\infty}$, indexed by all coarse time intervals (s,t) and satisfying $\mathcal{A}_{r,t} = \mathcal{A}_{r,s} \otimes \mathcal{A}_{s,t}$ whenever $r \leq s \leq t$, corresponds to a dyadic coarse factorization.

Example 3.6. Given a coarse time interval (s,t), we consider $f_{s,t} : \Omega[\text{all}] \to \mathbb{R}$,

$$f_{s,t}(\omega) = \frac{1}{\sqrt{i}} \sum_{k:s[i] \leq k/i < t[i]} \tau_{k/i}(\omega) \quad \text{for } \omega \in \Omega[i].$$

Only $s[\infty], t[\infty]$ matter, in the sense that

$$\int_{\Omega[i]} \frac{|\tilde{f}[i] - f[i]|}{1 + |\tilde{f}[i] - f[i]|} \, dP[i] \xrightarrow[i \to \infty]{} 0 \tag{3.4}$$

if $f = f_{s,t}$, and $\tilde{f} = f_{\tilde{s},\tilde{t}}$ is such a function built for a different coarse time interval $(\tilde{s}, \tilde{t})$ satisfying $\tilde{s}[\infty] = s[\infty]$, $\tilde{t}[\infty] = t[\infty]$. Moreover, $\|\tilde{f}[i] - f[i]\|_{L_2[i]} \to 0$ for $i \to \infty$. We choose a sequence of coarse time intervals, $(s_n, t_n)_{n=1}^{\infty}$, such that the sequence of their refinements, $(s_n[\infty], t_n[\infty])$ is dense among all (usual, not coarse) intervals. The sequence $(f_{s_n,t_n})_{n=1}^{\infty}$ satisfies the condition of Lemma 2.17 and therefore it generates a coarse σ-field $\mathcal{A}$. It is easy to see that $\mathcal{A}$ does not depend on the choice of (s_n, t_n). Clearly, the refinement of $f_{s,t}$ is the increment $B(t[\infty]) - B(s[\infty])$ of the usual Brownian motion $B(\cdot)$.

Given three coarse instants $r \leq s \leq t$, we have

$$f_{r,t} = f_{r,s} + f_{s,t}.$$

It shows that $f_{r,t}$ is coarsely measurable w.r.t. the product of two coarse σ-fields $\mathcal{A}_{r,s} \otimes \mathcal{A}_{s,t}$, which implies $\mathcal{A}_{r,t} = \mathcal{A}_{r,s} \otimes \mathcal{A}_{s,t}$. So, we have a dyadic coarse factorization. We may call it the Brownian coarse factorization.

Example 3.7. Let $f_{s,t}(\omega)$ be the same as in Example 3.6 and in addition,

$$g_{s,t}(\omega) = \frac{1}{\sqrt{i}} \sum_{k:s[i] \leq k/i < t[i]} (-1)^k \tau_{k/i}(\omega) \quad \text{for } \omega \in \Omega[i].$$

In the scaling limit we get two independent Brownian motions B_1, B_2; the refinement of $f_{s,t}$ is $B_1(t[\infty]) - B_1(s[\infty])$, the refinement of $g_{s,t}$ is $B_2(t[\infty]) - B_2(s[\infty])$. By the way, $(-1)^k$ cannot be replaced with $(-1)^{k-s[i]}$; it would violate the condition of Lemma 2.17.

We may also consider

$$f_{s,t}^{(n)}(\omega) = \frac{1}{\sqrt{i}} \sum_{k:s[i] \le k/i < t[i]} \exp\left(2\pi i \frac{k}{n}\right) \tau_{k/i}(\omega) \quad \text{for } \omega \in \Omega[i]$$

for $n = 1, 2, 3, \ldots$ (here $i = \sqrt{-1}$, while i is an integer). In the scaling limit we get two real-valued Brownian motions B_1, B_2 and infinitely many complex-valued Brownian motion $B_3, B_4, \ldots$ All B_n are independent.

Another construction of that kind:

$$f_{s,t}^{(\lambda)}(\omega) = \frac{1}{\sqrt{i}} \sum_{k:s[i] \le k/i < t[i]} \exp\left(2\pi i \lambda \frac{k}{\sqrt{i}}\right) \tau_{k/i}(\omega) \quad \text{for } \omega \in \Omega[i].$$

In the scaling limit, each $\lambda \in (0, \infty)$ gives a complex-valued Brownian motion B_λ. Any finite or countable set of numbers λ may be used, and leads to independent Brownian motions. Note that we cannot use more than a countable set of λ, since separability is stipulated by the definition of a coarse probability space.

Example 3.8. For $n = 1, 2, \ldots$ we introduce

$$f_{s,t}^{(n)}(\omega) = \frac{1}{\sqrt{i}} \sum_{k:s[i] \le k/i \le (k+n)/i < t[i]} \prod_{m=1}^{n} \tau_{(k+m)/i}(\omega) \quad \text{for } \omega \in \Omega[i].$$

In the scaling limit we get independent Brownian motions B_n.

Another construction of that kind:

$$f_{s,t}^{(\lambda)}(\omega) = \frac{1}{\sqrt{i}} \sum_{k:s[i] \le k/i \le (k+\lambda\sqrt{i})/i < t[i]} \prod_{m=1}^{\text{entier}(\lambda\sqrt{i})} \tau_{(k+m)/i}(\omega) \quad \text{for } \omega \in \Omega[i];$$

any finite or countable set of numbers $\lambda \in (0, \infty)$ may be used, and leads to independent Brownian motions B_λ.

Note that we cannot take the product over $m = 1, \ldots, \text{entier}(\lambda i)$; that would destroy factorizability.

Example 3.9. Here we restrict ourselves to $i \in \{2, 4, 8, 16, \ldots\}$, thus violating a little of our framework. We let for $\omega \in \Omega[i]$, $i = 2^n$,

$$g_{s,t}(\omega) = \sum_{k:s[i] \le k/i < (k+n-1)/i < t[i]} \frac{1 + \tau_{k/i}(\omega)}{2} \prod_{m=1}^{n-1} \frac{1 - \tau_{(k+m)/i}(\omega)}{2}.$$

That is, $g_{s,t} : \Omega[\text{all}] \to \{0, 1, 2, \ldots\}$ counts combinations '$+ - \ldots -$' of one plus sign and $(n-1)$ minus signs in succession. In the scaling limit we get the Poisson process.

Example 3.10. Let $f_{s,t}$ be as in Example 3.6 (Brownian), while $g_{s,t}$ is as in Example 3.9 (Poisson). Taken together, they generate a coarse σ-field. The corresponding scaling limit consists of two *independent* processes, Brownian and Poisson.

Let $\big((\Omega[i], \mathcal{F}[i], P[i])_{i=1}^{\infty}, \mathcal{A}\big)$ be a dyadic coarse factorization. Being a coarse probability space, it has a refinement $(\Omega, \mathcal{F}, P)$. For every coarse time interval (s,t) we have a coarse sub-σ-field $\mathcal{A}_{s,t} \subset \mathcal{A}$ and its refinement, a sub-σ-field $\mathcal{F}_{s,t}[\infty] \subset \mathcal{F}$. By Lemma 3.1,

$$\mathcal{F}_{r,t}[\infty] = \mathcal{F}_{r,s}[\infty] \otimes \mathcal{F}_{s,t}[\infty] \quad \text{whenever } r \leq s \leq t.$$

Lemma 3.11. *If $s[\infty] = t[\infty]$ then $\mathcal{F}_{s,t}[\infty]$ is degenerate (that is, contains sets of probability 0 or 1 only).*

Proof. Consider the coarse instant r,

$$r[i] = \begin{cases} s[i] & \text{for } i \text{ even,} \\ t[i] & \text{for } i \text{ odd.} \end{cases}$$

For every $A \in \mathcal{A}_{s,r}$,

$$P(A[\infty]) = \lim_{i \to \infty} P[i]\big(A[i]\big) = \lim_{i \to \infty} P[2i]\big(A[2i]\big) \in \{0,1\},$$

since $\mathcal{A}_{s,r}[2i]$ is degenerate. So, $\mathcal{F}_{s,r}[\infty]$ is degenerate. Similarly, $\mathcal{F}_{r,t}[\infty]$ is degenerate. However, $\mathcal{F}_{s,t}[\infty] = \mathcal{F}_{s,r}[\infty] \otimes \mathcal{F}_{r,t}[\infty]$. $\qquad\square$

Lemma 3.12. *$\mathcal{F}_{s,t}[\infty]$ depends only on $s[\infty], t[\infty]$.*

Proof. Let (u,v) be another coarse time interval such that $u[\infty] = s[\infty]$ and $v[\infty] = t[\infty]$; we have to prove that $\mathcal{F}_{s,t}[\infty] = \mathcal{F}_{u,v}[\infty]$. Assume that $s[\infty] < t[\infty]$ (otherwise both $\mathcal{F}_{s,t}[\infty]$ and $\mathcal{F}_{u,v}[\infty]$ are degenerate). Assume also that $s[i] \leq v[i]$ and $u[i] \leq t[i]$ for all i (otherwise we correct them on a finite set of indices i).

Further, we may assume that $s \leq u \leq v \leq t$; otherwise we turn to $s \wedge u \leq s \vee u \leq t \wedge v \leq t \vee v$, where $(s \wedge u)[i] = s[i] \wedge u[i] = \min\big(s[i], u[i]\big)$, etc. Both $\mathcal{F}_{s,t}[\infty]$ and $\mathcal{F}_{u,v}[\infty]$ are sandwiched between $\mathcal{F}_{s \wedge u, t \vee v}[\infty]$ and $\mathcal{F}_{s \vee u, t \wedge v}[\infty]$.

Finally, $\mathcal{F}_{s,t}[\infty] = \mathcal{F}_{s,u}[\infty] \otimes \mathcal{F}_{u,v}[\infty] \otimes \mathcal{F}_{v,t}[\infty] = \mathcal{F}_{u,v}[\infty]$, since $\mathcal{F}_{s,u}[\infty]$ and $\mathcal{F}_{v,t}[\infty]$ are degenerate by Lemma 3.11. $\qquad\square$

So, a sub-σ-field $\mathcal{F}_{s,t} \subset \mathcal{F}$ is well-defined for every interval $(s,t) \subset \mathbb{R}$ (rather than a coarse time interval), and

$$\mathcal{F}_{r,t} = \mathcal{F}_{r,s} \otimes \mathcal{F}_{s,t} \quad \text{whenever } -\infty < r \leq s \leq t < +\infty.$$

Lemma 3.13. *The union of sub-σ-fields $\mathcal{F}_{s+\varepsilon, t-\varepsilon}$ over $\varepsilon > 0$ generates $\mathcal{F}_{s,t}$.*

Proof. Consider $\mathcal{F}_{\varepsilon,1}$. We have to prove that $\mathbb{E}\left(x \mid \mathcal{F}_{\varepsilon,1}\right)$ converges to x (in $L_2(\Omega)$, for $\varepsilon \to 0+$) for every $x \in L_2(\mathcal{F}_{0,1})$, or for $x[\infty]$ where $x \in L_2(\mathcal{A}_{0,1})$. Assume the contrary. Then

$$\|\mathbb{E}\left(x[\infty] \mid \mathcal{F}_{\varepsilon,1}\right)\| < c < \|x[\infty]\|$$

for all ε small enough, and some constant c. We know that

$$\mathbb{E}\left(x[\infty] \mid \mathcal{F}_{\varepsilon,1}\right) = \operatorname{Lim}\mathbb{E}\left(x[i] \mid \mathcal{F}_{\varepsilon,1}[i]\right)$$

for each ε.[17] Therefore

$$\|\mathbb{E}\left(x[i] \mid \mathcal{F}_{\varepsilon,1}[i]\right)\| \xrightarrow[i\to\infty]{} \|\mathbb{E}\left(x[\infty] \mid \mathcal{F}_{\varepsilon,1}\right)\| < c.$$

We choose a sequence $\varepsilon[i] \xrightarrow[i\to\infty]{} 0$ such that $\|\mathbb{E}\left(x[i] \mid \mathcal{F}_{\varepsilon[i],1}[i]\right)\| < c$ for all i large enough. However, $\operatorname{Lim}\mathbb{E}\left(x[i] \mid \mathcal{F}_{\varepsilon[i],1}[i]\right) = \mathbb{E}\left(x[\infty] \mid \mathcal{F}_{\varepsilon[\infty],1}\right) = \mathbb{E}\left(x[\infty] \mid \mathcal{F}_{0,1}\right) = x[\infty]$; a contradiction. $\qquad\square$

3.3 Scaling Limit of Fourier-Walsh Coefficients

We still consider a dyadic coarse factorization. The Hilbert space $L_2[i] = L_2\left(\Omega[i], \mathcal{F}[i], P[i]\right)$ consists of all functions of random signs τ_m, $m \in \frac{1}{i}\mathbb{Z}$. The well-known Fourier-Walsh orthonormal basis of $L_2[i]$ consists of products

$$\tau_M = \prod_{m\in M} \tau_m, \quad M \in \mathcal{C}[i], \quad \mathcal{C}[i] = \{M \subset \tfrac{1}{i}\mathbb{Z} : M \text{ is finite}\}.$$

Every $f \in L_2[i]$ is of the form

$$f = \sum_M \hat{f}_M \tau_M = \hat{f}_\emptyset + \sum_{m\in\frac{1}{i}\mathbb{Z}} \hat{f}_{\{m\}}\tau_m + \sum_{m_1,m_2\in\frac{1}{i}\mathbb{Z}, m_1<m_2} \hat{f}_{\{m_1,m_2\}}\tau_{m_1}\tau_{m_2} + \dots;$$

coefficients $\hat{f}_M$ are called Fourier-Walsh coefficients of f. We define the *spectral measure* μ_f on the countable set $\mathcal{C}[i]$ by

$$\mu_f(\mathcal{M}) = \sum_{M\in\mathcal{M}} |\hat{f}_M|^2 \quad \text{for } \mathcal{M} \subset \mathcal{C}[i];$$

it is a finite positive measure,

$$\mu_f(\mathcal{C}[i]) = \|f\|^2; \quad \mu_f(\{\emptyset\}) = (\mathbb{E}\,f)^2; \quad \mu_f(\mathcal{C}[i] \setminus \{\emptyset\}) = \operatorname{Var}(f).$$

Let (s,t) be a coarse time interval. We have

---

[17] Or rather, an appropriate coarse instant is meant in $\mathcal{F}_{\varepsilon,1}[i]$.

$$\mathbb{E}\left(\tau_M \,\middle|\, \mathcal{F}_{s,t}[i]\right) = \begin{cases} \tau_M & \text{if } M \subset [s[i], t[i]), \\ 0 & \text{otherwise;} \end{cases}$$

$$\|\mathbb{E}\left(f \,\middle|\, \mathcal{F}_{s,t}[i]\right)\|^2 = \mu_f\left(\{M \in \mathcal{C}[i] : M \subset [s[i], t[i])\}\right).$$

We apply it to $f = x[i]$ for an arbitrary $x \in L_2(\mathcal{A})$ and arbitrary i; μ_f becomes $\mu_{x[i]}$ or $\mu_x[i]$; by (3.3),

$$\mu_x[i]\left(\{M \in \mathcal{C}[i] : M \subset [s[i], t[i])\}\right) = \|\mathbb{E}\left(x[i] \,\middle|\, \mathcal{F}_{s,t}[i]\right)\|^2$$
$$\xrightarrow[i\to\infty]{} \|\mathbb{E}\left(x[\infty] \,\middle|\, \mathcal{F}_{s,t}[\infty]\right)\|^2.$$

For every $\varepsilon > 0$ we can choose s, t so that $\|x[\infty]\|^2 - \|\mathbb{E}\left(x[\infty] \,\middle|\, \mathcal{F}_{s,t}[\infty]\right)\|^2 \le \varepsilon$, and moreover,

$$\mu_x[i]\left(\{M \in \mathcal{C}[i] : M \subset [s[i], t[i])\}\right) \le \varepsilon \quad \text{for all } i. \tag{3.5}$$

We consider each $\mu_x[i]$ as a measure on the space $\mathcal{C}[\infty]$ of all compact subsets of $\mathbb{R}$, equipped with the Hausdorff metric; the metric is

$$\operatorname{dist}(M_1, M_2) = \sup_{x \in \mathbb{R}} \left| \min_{y \in M_1} |x - y| - \min_{y \in M_2} |x - y| \right| \tag{3.6}$$

for nonempty M_1, M_2, and $\operatorname{dist}(\emptyset, M) = 1$ for $M \ne \emptyset$. Clearly, $\mathcal{C}[i] \subset \mathcal{C}[\infty]$ for each i; thus, a measure on $\mathcal{C}[i]$ is also a measure on $\mathcal{C}[\infty]$.[18] The set $\{M \in \mathcal{C}[\infty] : M \subset [u, v]\}$ is well-known to be compact, for every $[u, v] \subset \mathbb{R}$. Thus, (3.5) shows that the sequence of measures $\mu_x[i]$ on $\mathcal{C}[\infty]$ is tight.

Let (s_1, t_1) and (s_2, t_2) be two coarse time intervals, $s_1 \le t_1 \le s_2 \le t_2$. Sub-$\sigma$-fields $\mathcal{F}_{s_1,t_1}[i]$ and $\mathcal{F}_{s_2,t_2}[i]$ are independent; they generate a sub-σ-field that may be denoted by

$$\mathcal{F}_{(s_1,t_1)\cup(s_2,t_2)}[i] = \mathcal{F}_{s_1,t_1}[i] \otimes \mathcal{F}_{s_2,t_2}[i].$$

We have

$$\mathbb{E}\left(\tau_M \,\middle|\, \mathcal{F}_{(s_1,t_1)\cup(s_2,t_2)}[i]\right) = \begin{cases} \tau_M & \text{if } M \subset [s_1[i], t_1[i]) \cup [s_2[i], t_2[i]), \\ 0 & \text{otherwise;} \end{cases}$$

$$\|\mathbb{E}\left(f \,\middle|\, \mathcal{F}_{(s_1,t_1)\cup(s_2,t_2)}[i]\right)\|^2$$
$$= \mu_f\left(\{M \in \mathcal{C}[i] : M \subset [s_1[i], t_1[i]) \cup [s_2[i], t_2[i])\}\right);$$
$$\mu_x[i]\left(\{M \in \mathcal{C}[i] : M \subset [s_1[i], t_1[i]) \cup [s_2[i], t_2[i])\}\right)$$
$$= \|\mathbb{E}\left(x[i] \,\middle|\, \mathcal{F}_{(s_1,t_1)\cup(s_2,t_2)}[i]\right)\|^2 \xrightarrow[i\to\infty]{} \|\mathbb{E}\left(x[\infty] \,\middle|\, \mathcal{F}_{(s_1,t_1)\cup(s_2,t_2)}[\infty]\right)\|^2,$$

---

[18] One may turn $(\mathcal{C}[i])_{i=1}^{\infty}$ into a coarse Polish space, and identify its refinement with $\mathcal{C}[\infty]$. It leads to a joint compactification of all $\mathcal{C}[i]$ and $\mathcal{C}[\infty]$, which is a suitable framework for weak convergence of measures on $\mathcal{C}[i]$ to a measure on $\mathcal{C}[\infty]$. However, it is simpler to use natural embeddings, $\mathcal{C}[i] \subset \mathcal{C}[\infty]$.

where $\mathcal{F}_{(s_1,t_1)\cup(s_2,t_2)}[\infty] = \mathcal{F}_{s_1,t_1}[\infty] \otimes \mathcal{F}_{s_2,t_2}[\infty] = \mathcal{F}_{s_1[\infty],t_1[\infty]} \otimes \mathcal{F}_{s_2[\infty],t_2[\infty]}$.
A generalization of (3.3) to the product of more than two spaces was used here.

The same holds for more than two coarse time intervals:

$$\mu_x[i](\{M \in \mathcal{C}[i] : M \subset [s_1[i],t_1[i]) \cup \ldots \cup [s_n[i],t_n[i])\})$$
$$\xrightarrow[i\to\infty]{} \|\mathbb{E}\left(x[\infty] \,|\, \mathcal{F}_{(s_1,t_1)\cup\ldots\cup(s_n,t_n)}[\infty]\right)\|^2 . \quad (3.7)$$

We have convergence of spectral measures on a special class of subsets of $\mathcal{C}[\infty]$. Note that the intersection of two such subsets is again such a subset. Therefore, the convergence holds on the algebra of subsets generated by the class. A generic element of the algebra is the union of a finite number of 'cells' of the form

$$\{M \in \mathcal{C}[\infty] : M \subset \cup_{k=1}^n [s_k,t_k) \text{ and } M \cap [s_k,t_k) \neq \emptyset \text{ for } k = 1,\ldots,n\}; \quad (3.8)$$

here $[s_k,t_k) \subset \mathbb{R}$ are usual (rather than coarse) time intervals. (Endpoints may be neglected, as we will see soon.) The diameter of the cell (3.8) (w.r.t. the metric (3.6)) does not exceed $\max_k(t_k - s_k)$. Thus, we get weak convergence of measures, which proves the following result.

Theorem 3.14. *For every dyadic coarse factorization* $((\Omega[i], \mathcal{F}[i], P[i])_{i=1}^\infty, \mathcal{A})$ *and every* $x \in L_2(\mathcal{A})$, *the sequence* $(\mu_x[i])_{i=1}^\infty$ *of spectral measures converges weakly to a (finite, positive) measure* $\mu_x[\infty]$ *on the Polish space* $\mathcal{C}[\infty]$.

Convergence of measures $\mu_x[i]$ on a 'cell' of the form (3.7) (or (3.8)) does not ensure that the limit is $\mu_x[\infty]$ on the 'cell'.[19] Rather, the limit lies between $\mu_x[\infty]$-measures of the interior and the closure of the cell,

$$\mu_x[\infty](\{M \in \mathcal{C}[\infty] : M \subset (s_1,t_1) \cup \ldots \cup (s_n,t_n)\})$$
$$\leq \|\mathbb{E}\left(x[\infty] \,|\, \mathcal{F}_{(s_1,t_1)\cup\ldots\cup(s_n,t_n)}\right)\|^2$$
$$\leq \mu_x[\infty](\{M \in \mathcal{C}[\infty] : M \subset [s_1,t_1] \cup \ldots \cup [s_n,t_n]\}). \quad (3.9)$$

Lemma 3.15. *For every* $t \in \mathbb{R}$,

$$\mu_x[\infty](\{M \in \mathcal{C}[\infty] : M \ni t\}) = 0.$$

Proof. Lemma 3.13 gives us

$$\|\mathbb{E}\left(x[\infty] \,|\, \mathcal{F}_{(-\infty,-\varepsilon)\cup(\varepsilon,+\infty)}\right)\|^2 \xrightarrow[\varepsilon\to 0]{} \|x[\infty]\|^2 ;$$

therefore

$$\mu_x[\infty](\{M \in \mathcal{C}[\infty] : M \subset (-\infty,\varepsilon] \cup [\varepsilon,+\infty)\}) \xrightarrow[\varepsilon\to 0]{} \mu_x[\infty](\mathcal{C}[\infty]).$$

$\square$

---

[19] Think for example about an atom at the point $\frac{1}{n}$ of $\mathbb{R}$, and 'cells' of the form $(x,y]$.

Applying Fubini's theorem we see that $\mu_x[\infty]$ is concentrated on (the set of all) compact sets M of Lebesgue measure 0 (therefore, nowhere dense).

Due to Lemma 3.15 we see that the boundary of a 'cell' is negligible (of measure 0); inequalities (3.9) are, in fact, equalities. So,

$$\mu_x[\infty](\{M \in \mathcal{C}[\infty] : M \subset E\}) = \|\mathbb{E}\left(x[\infty] \,|\, \mathcal{F}_E\right)\|^2, \qquad (3.10)$$

where $E \subset \mathbb{R}$ is an arbitrary elementary set, that is, a finite union of intervals (treated modulo finite sets), $E = (s_1, t_1) \cup \ldots \cup (s_n, t_n)$, and $\mathcal{F}_E = \mathcal{F}_{s_1,t_1} \otimes \cdots \otimes \mathcal{F}_{s_n,t_n}$.

For a finite i, the Fourier-Walsh basis decomposes $L_2[i]$ into one-dimensional subspaces indexed by $M \in \mathcal{C}[i]$, and each subset $\mathcal{M} \subset \mathcal{C}[i]$ leads to a subspace $H_\mathcal{M}$ of $L_2[i]$ spanned by τ_M, $M \in \mathcal{M}$. In particular, for a subset of the form $\mathcal{M}_E = \{M \in \mathcal{C}[i] : M \subset E\}$ we have $H_{\mathcal{M}_E} = L_2(\Omega[i], \mathcal{F}_E[i], P[i])$.

Similarly, for the limiting object, the subspace $H_{\mathcal{M}_E} = L_2(\Omega, \mathcal{F}_E, P)$ of $L_2[\infty]$ corresponds to the set $\mathcal{M}_E = \{M \in \mathcal{C}[\infty] : M \subset E\}$. In Sect. 3.4 a subspace $H_\mathcal{M} \subset L_2[\infty]$ will be defined for every Borel set $\mathcal{M} \subset \mathcal{C}[\infty]$.

3.4 The Limiting Object

Definition 3.16. *A* continuous factorization (*of probability spaces, over* $\mathbb{R}$) *consists of a probability space* $(\Omega, \mathcal{F}, P)$ *and a two-parameter family* $(\mathcal{F}_{s,t})_{s \leq t}$ *of sub-σ-fields* $\mathcal{F}_{s,t} \subset \mathcal{F}$ *such that*[20]

(a) $$\mathcal{F}_{r,t} = \mathcal{F}_{r,s} \otimes \mathcal{F}_{s,t} \quad \text{whenever } r \leq s \leq t$$

(*that is,* $\mathcal{F}_{r,s}$ *and* $\mathcal{F}_{s,t}$ *are independent, and together generate* $\mathcal{F}_{r,t}$),

(b) $$\bigcup_{\varepsilon > 0} \mathcal{F}_{s+\varepsilon, t-\varepsilon} \text{ generates } \mathcal{F}_{s,t} \text{ whenever } s < t,$$

and

(c) $$\bigcup_{n=1}^{\infty} \mathcal{F}_{-n,n} \text{ generates } \mathcal{F}.$$

The refinement of any dyadic coarse factorization is a continuous factorization (as was shown in Sect. 3.2).

Definition 3.17. *Let* $\left((\Omega, \mathcal{F}, P), (\mathcal{F}_{s,t})_{s \leq t}\right)$ *be a continuous factorization, and* $x \in L_2(\Omega, \mathcal{F}, P)$. *The* spectral measure μ_x *of* x *is the* (*finite, positive*) *measure on the space* $\mathcal{C} = \mathcal{C}[\infty]$ *of compact subsets of* $\mathbb{R}$ *such that*

$$\mu_x(\{M \in \mathcal{C} : M \subset E\}) = \|\mathbb{E}\left(x \,|\, \mathcal{F}_E\right)\|^2$$

for all elementary sets $E \subset \mathbb{R}$.

---

[20] Here r, s, t are real numbers; coarse instants are not used in Sects. 3.4, 3.5.

Uniqueness of μ_x is checked easily. Existence of μ_x is proven in Sect. 3.3 by discrete approximation, assuming that the continuous factorization is the refinement of a dyadic coarse factorization. Another proof, without approximation, will be given by Lemma 3.23.

The spectral measure is concentrated on (the set of all) nowhere dense compact sets, and

$$\mu_x(\{M \in \mathcal{C} : M \ni t\}) = 0 \quad \text{for each } t \in \mathbb{R}, \tag{3.11}$$

which follows from Lemma 3.20 for $s = t$, since $\mathcal{F}_{t,t} = \mathcal{F}_{t,t} \otimes \mathcal{F}_{t,t}$ is degenerate.

Example 3.18. The refinement of the Brownian coarse factorization (see Example 3.6) is the Brownian continuous factorization,

$$\mathcal{F}_{s,t} \text{ is generated by } \{B(v) - B(u) : s \le u \le v \le t\},$$

where $B(\cdot)$ is the usual Brownian motion. Every $x \in L_2$ admits Itô's decomposition into multiple stochastic integrals,

$$x = \hat{x}(\emptyset) + \int \hat{x}(\{t_1\})\, \mathrm{d}B(t_1) + \iint_{t_1 < t_2} \hat{x}(\{t_1, t_2\})\, \mathrm{d}B(t_1)\mathrm{d}B(t_2) + \dots$$

$$= \sum_{n=0}^{\infty} \int_{t_1 < \dots < t_n} \dots \int \hat{x}(\{t_1, \dots, t_n\})\, \mathrm{d}B(t_1) \dots \mathrm{d}B(t_n),$$

where $\hat{x} \in L_2(\mathcal{C}_{\text{finite}})$, $\mathcal{C}_{\text{finite}}$ being the space of all finite subsets of $\mathbb{R}$, equipped with the natural (Lebesgue) measure, making the transform $x \leftrightarrow \hat{x}$ unitary, according to the formula

$$\mathbb{E}\,|x|^2 = |\hat{x}(\emptyset)|^2 + \int |\hat{x}(\{t_1\})|^2\, \mathrm{d}t_1 + \iint_{t_1 < t_2} |\hat{x}(\{t_1, t_2\})|^2\, \mathrm{d}t_1\mathrm{d}t_2 + \dots$$

$$= \sum_{n=0}^{\infty} \int_{t_1 < \dots < t_n} \dots \int |\hat{x}(\{t_1, \dots, t_n\})|^2\, \mathrm{d}t_1 \dots \mathrm{d}t_n.$$

The spectral measure μ_x of x is

$$\mu_x(A) = \sum_{n=0}^{\infty} \int_{t_1 < \dots < t_n, \{t_1, \dots, t_n\} \in A} \dots \int |\hat{x}(\{t_1, \dots, t_n\})|^2\, \mathrm{d}t_1 \dots \mathrm{d}t_n.$$

This is an important property of the Brownian continuous factorization: the spectral measure (of any random variable) is concentrated on the subset $\mathcal{C}_{\text{finite}} \subset \mathcal{C}$, and absolutely continuous w.r.t. the Lebesgue measure on $\mathcal{C}_{\text{finite}}$.

In particular, for $x = \exp(i\sqrt{\lambda}B(t))$ the measure μ_x is just the distribution of the Poisson process of rate λ on $(0, t)$. Indeed,

$$\exp\bigl(i\sqrt{\lambda}B(t)\bigr) = e^{-\lambda t/2} \sum_{n=0}^{\infty} \lambda^{n/2} \underset{0<t_1<\cdots<t_n<t}{\int \cdots \int} dB(t_1) \ldots dB(t_n).$$

Example 3.19. Recall the process Y_ε of Example 1.2;

$$Y_\varepsilon(t) = \exp\bigl(iB(\ln t) - iB(\ln\varepsilon)\bigr).$$

We define $\mathcal{F}_{s,t}$ as the σ-field generated by 'multiplicative increments' $Y_\varepsilon(v)/Y_\varepsilon(u)$ for all $(u,v) \subset (s,t)$, that is, by (usual) Brownian increments on $(\ln s, \ln t)$. The spectral measure $\mu_{Y_\varepsilon(t)}$ is the distribution of a non-homogeneous Poisson process on (ε, t), the image of the usual Poisson process (of rate 1) on $(\ln\varepsilon, \ln t)$ under the time change $u \mapsto e^u$. The rate of the non-homogeneous Poisson process is $\lambda(s) = 1/s$.

The limiting process Y was discussed in Example 1.2. It may be treated as the refinement of Y_ε for $\varepsilon \to 0$ (I leave the details to the reader). The spectral measure $\mu_{Y(t)}$ should be the distribution of a non-homogeneous Poisson process on $(0,t)$, at the rate $\lambda(s) = 1/s$. Random points accumulate to 0; we add 0 to the random set, making it compact. However, the equality $\mu(\{M : M \ni 0\}) = 1$ does not conform to Lemma 3.15! It happens because the limiting object is not a *continuous* factorization. Denote by $\mathcal{F}_{0+,1}$ the σ-field generated by $\cup_{\varepsilon>0}\mathcal{F}_{\varepsilon,1}$. Every $Y(1)/Y(t)$ for $t > 0$ is $\mathcal{F}_{0+,1}$-measurable, but $Y(1)$ is not. The global phase is missing. Of course, for every $t > 0$, there exists an independent complement of $\mathcal{F}_{0+,t}$ in $\mathcal{F}_{-\infty,t}$ (for example, the σ-field generated by $Y(t)$). However, we cannot choose a single complement (to be denoted by $\mathcal{F}_{-\infty,0+}$) for all $t > 0$, since the tail σ-field $\cap_{t>0}\mathcal{F}_{-\infty,t}$ is degenerate.

Lemma 3.20. *For every continuous factorization* $\bigl((\Omega,\mathcal{F},P),(\mathcal{F}_{s,t})_{s\leq t}\bigr)$ *and every* $s \leq t$,

$$\mathcal{F}_{s,t} = \bigcap_{\varepsilon>0} \mathcal{F}_{s-\varepsilon,t+\varepsilon}.$$

Proof. The σ-field $\cap_{\varepsilon>0}\mathcal{F}_{0,\varepsilon}$ is degenerate by Kolmogorov's zero-one law applied to $\mathcal{F}_{1,\infty}, \mathcal{F}_{1/2,1}, \mathcal{F}_{1/3,1/2}, \ldots$ Further, $\mathcal{F}_{-\infty,\varepsilon} = \mathcal{F}_{-\infty,0} \otimes \mathcal{F}_{0,\varepsilon} \xrightarrow[\varepsilon\to0]{} \mathcal{F}_{-\infty,0}$. Though the equality $\lim(\mathcal{A} \vee \mathcal{B}_n) = \mathcal{A} \vee (\lim\mathcal{B}_n)$ does not hold in general, it does hold for independent $\mathcal{A}$ and $\mathcal{B}_1$ ($\mathcal{B}_1 \supset \mathcal{B}_2 \supset \ldots$), which is a rather trivial part of Weizsäcker's criteria [27]. The rest of the proof is left to the reader. $\qquad\square$

The theory of direct integrals of Hilbert spaces may be used on the way to Theorem 3.26. In fact, I did so in [15, Th. 2.3]. Here, however, I choose a self-contained presentation. First, a general result of measure theory, useful for proving the existence of μ_x (without dyadic approximation).

Lemma 3.21. *Let X be a compact topological space, $\mathcal{A}$ an algebra of subsets of X, and $\mu : \mathcal{A} \to [0,\infty)$ an additive function satisfying the following regularity condition:*

For every $A \in \mathcal{A}$ and $\varepsilon > 0$ there exists $B \in \mathcal{A}$ such that $\overline{B} \subset A$ (here $\overline{B}$ is the closure of B) and $\mu(B) \geq \mu(A) - \varepsilon$.

Then μ has a unique extension to a measure on the σ-field generated by $\mathcal{A}$.

Proof. Due to a well-known theorem, it is enough to prove that μ is σ-additive on $\mathcal{A}$. Let $A_1 \supset A_2 \supset \dots$, $A_1, A_2, \dots \in \mathcal{A}$, $\cap A_k = \emptyset$; we have to prove that $\mu(A_k) \to 0$. Given $\varepsilon > 0$, we can choose $B_k \in \mathcal{A}$ such that $\overline{B}_k \subset A_k$ and $\mu(B_k) \geq \mu(A_k) - 2^{-k}\varepsilon$. Due to compactness, the relation $\cap \overline{B}_k \subset \cap A_k = \emptyset$ implies $\overline{B}_1 \cap \dots \cap \overline{B}_n = \emptyset$ for some n. Thus, $\mu(A_n) = \mu(A_1 \cap \dots \cap A_n) \leq \mu(B_1 \cap \dots \cap B_n) + \mu(A_1 \setminus B_1) + \dots + \mu(A_n \setminus B_n) < \varepsilon$. $\qquad\square$

Remark 3.22. All $A \in \mathcal{A}$ such that A and $X \setminus A$ both satisfy the regularity condition, are a subalgebra of $\mathcal{A}$. (The proof is left to the reader.) Therefore it is enough to check the condition for A and $X \setminus A$ where A runs over a set that generates the algebra $\mathcal{A}$.

Lemma 3.23. *The spectral measure μ_x exists for every $x \in L_2(\Omega, \mathcal{F}, P)$ and every continuous factorization $(\mathcal{F}_{s,t})_{s \leq t}$.*

Proof. First, compactness. We have $\|\mathbb{E}(x \mid \mathcal{F}_{-m,m})\|^2 \to \|x\|^2$ for $m \to \infty$ by 3.16(c); thus we may restrict ourselves to x measurable w.r.t. $\mathcal{F}_{-m,m}$ for some m. The corresponding part $\mathcal{C}_m = \{M \in \mathcal{C} : M \subset [-m, m]\}$ of $\mathcal{C}$ is compact.

Second, additivity on an algebra. We have an algebra $\mathcal{A}$ of subsets of $\mathcal{C}_m$, generated by 'cells' of the form (3.8). Such a cell leads to a subspace of $L_2(\Omega, \mathcal{F}_{-m,m}, P)$ spanned by products $f_1 \dots f_n$ where each f_k is measurable w.r.t. $\mathcal{F}_{s_k,t_k}$, square integrable, and $\mathbb{E} f_k = 0$. A partition of the interval $[-m, m]$ into n subintervals leads to a partition of $\mathcal{C}_m$ into 2^n parts, and a decomposition of $L_2(\Omega, \mathcal{F}_{-m,m}, P)$ into 2^n orthogonal subspaces. Thus, x decomposes into 2^n orthogonal vectors; their squared norms give us μ_x on a finite subalgebra (of cardinality 2^{2^n}) of $\mathcal{A}$. We see that μ_x is additive on such subalgebras. Their union (over all partitions of $[-m, m]$) is the whole $\mathcal{A}$, and any two of them are contained in some third; therefore, μ_x is additive on $\mathcal{A}$.

Third, regularity (required by Lemma 3.21). Due to Remark 3.22, regularity may be checked only for sets $A_E = \{M \in \mathcal{C}_m : M \subset E\}$ and $\mathcal{C}_m \setminus A_E$. It follows easily from 3.16(b) and Lemma 3.20. $\qquad\square$

Remark 3.24. In the proof of Lemma 3.23, an *orthogonal decomposition* of the Hilbert space $H = L_2(\Omega, \mathcal{F}, P)$ over the algebra $\mathcal{A}$ is constructed; that is, a family $(H_A)_{A \in \mathcal{A}}$ of (closed linear) subspaces $H_A \subset H$ such that $H_{A \cup B} = H_A \oplus H_B$ (it means that H_A and H_B are orthogonal, and their sum is $H_{A \cup B}$) whenever $A \cap B = \emptyset$, and $H_{\mathcal{C}} = H$. The decomposition satisfies

$$H_{\mathcal{M}_E} = L_2(\Omega, \mathcal{F}_E, P),$$

where $\mathcal{M}_E = \{M \in \mathcal{C} : M \subset E\}$, and is uniquely determined by this property.

The following general result will help us construct $H_\mathcal{M}$ for all Borel sets $\mathcal{M} \subset \mathcal{C}$.

Lemma 3.25. *Let X be a set, $\mathcal{A}$ an algebra of subsets of X, H a Hilbert space, and $(H_A)_{A \in \mathcal{A}}$ an orthogonal decomposition of H over $\mathcal{A}$. Assume that for every $x \in H$ the additive function[21] $A \mapsto \| \text{Proj}_{H_A} x \|^2$ on $\mathcal{A}$ can be extended to a measure on the σ-field $\sigma(\mathcal{A})$ generated by $\mathcal{A}$. Then the orthogonal decomposition can be extended to an orthogonal decomposition $(H_B)_{B \in \sigma(\mathcal{A})}$, σ-additive in the sense that[22] $H_{B_1 \cup B_2 \cup \ldots} = H_{B_1} \oplus H_{B_2} \oplus \ldots$ whenever $B_1, B_2, \cdots \in \sigma(\mathcal{A})$ are pairwise disjoint.*

Proof. The extension of the additive function $\mu_x : \mathcal{A} \to [0, \infty)$, $\mu_x(A) = \| \text{Proj}_{H_A} x \|^2$, to a measure on $\sigma(\mathcal{A})$ is unique; denote it by μ_x again. Consider the set of all $B \in \sigma(\mathcal{A})$ such that there exists a subspace $H_B \subset H$ satisfying $\| \text{Proj}_{H_B} x \|^2 = \mu_x(B)$ for all $x \in H$. The set contains $\mathcal{A}$, and is a monotone class (that is, closed under the limit of monotone sequences), which is easy to check. Therefore the set is the whole $\sigma(\mathcal{A})$. $\qquad \square$

Combining Lemmas 3.23 and 3.25 we conclude.

Theorem 3.26. *For every continuous factorization $((\Omega, \mathcal{F}, P), (\mathcal{F}_{s,t})_{s \le t})$ there exists one and only one σ-additive orthogonal decomposition $(H_\mathcal{M})$ of the Hilbert space $L_2(\Omega, \mathcal{F}, P)$ over the Borel σ-field of the space $\mathcal{C}$ (of compact subsets of $\mathbb{R}$) such that $H_{\mathcal{M}_E} = L_2(\Omega, \mathcal{F}_E, P)$ for every elementary set $E \subset \mathbb{R}$ (that is, a finite union of intervals); here $\mathcal{M}_E = \{M \in \mathcal{C} : M \subset E\}$. The orthogonal decomposition is related to spectral measures by*

$$\| \text{Proj}_{H_\mathcal{M}} f \|^2 = \mu_f(\mathcal{M}) \qquad (3.12)$$

for all $f \in L_2(\Omega, \mathcal{F}, P)$ and all Borel sets $\mathcal{M} \subset \mathcal{C}$.

3.5 Time Shift; Noise

Let $((\Omega[i], \mathcal{F}[i], P[i])_{i=1}^\infty, \mathcal{A})$ be a dyadic coarse factorization. For each i the lattice $\frac{1}{i}\mathbb{Z}$ acts on $\Omega[i]$ by measure preserving transformations $\alpha_t : \Omega[i] \to \Omega[i]$ (time shift),

$$\alpha_t(\omega)(s) = \omega(s - t) \quad \text{for all } s \in \frac{1}{i}\mathbb{Z}.$$

For each coarse instant $t = (t[i])_{i=1}^\infty$ we have a map $\alpha_t : \Omega[\text{all}] \to \Omega[\text{all}]$,

$$\alpha_t(\omega)[i](s) = \omega[i](s - t[i]) \quad \text{for all } s \in \frac{1}{i}\mathbb{Z}.$$

Such α_t is an automorphism of the dyadic coarse sample space, but the coarse σ-field $\mathcal{A}$ need not be invariant under α_t. We consider such a condition:

---

[21] Here Proj_{H_A} is the orthogonal projection $H \to H_A$.
[22] That is, $H_{B_1 \cup B_2 \cup \ldots}$ is the closure of the algebraic sum of H_{B_k}.

$$\mathcal{A} \text{ is invariant under } \alpha_t \text{ for every coarse instant } t. \qquad (3.13)$$

Dyadic coarse factorizations of Examples 3.6, 3.8, 3.9, 3.10 satisfy (3.13), but that of Example 3.7 does not.

If (3.13) is satisfied, then the refinement $\alpha_t[\infty] = \text{Lim}_{i\to\infty,\mathcal{A}} \alpha_t[i]$ is an automorphism of the refinement $(\Omega, \mathcal{F}, P)$ of the dyadic coarse factorization. Existence of the limit for *every* converging sequence $t = (t[i])$ implies that $\alpha_t[\infty]$ depends on $t[\infty]$ only (see Lemma 3.29 below), and we get a one-parameter group $(\alpha_t)_{t\in\mathbb{R}}$ of automorphisms (that is, invertible measure preserving transformations mod 0) of $(\Omega, \mathcal{F}, P)$. The group is continuous in the sense that $\mathbb{P}\left(A \triangle \alpha_t(A)\right) \xrightarrow[t\to 0]{} 0$ for all $A \in \mathcal{F}$, which is ensured by (3.13) (see Lemma 3.29 again).

Definition 3.27. *A* noise $\left((\Omega, \mathcal{F}, P), (\mathcal{F}_{s,t})_{s\le t}, (\alpha_t)_{t\in\mathbb{R}}\right)$ *consists of a continuous factorization* $\left((\Omega, \mathcal{F}, P), (\mathcal{F}_{s,t})_{s\le t}\right)$ *and a one-parameter group of automorphisms* α_t *of* $(\Omega, \mathcal{F}, P)$ *such that*

$$\alpha_t^{-1}(\mathcal{F}_{r,s}) = \mathcal{F}_{r-t,s-t} \quad \text{for all } r, s, t \in \mathbb{R}, \ r \le s,$$
$$P\left(A \triangle \alpha_t^{-1}(A)\right) \xrightarrow[t\to 0]{} 0 \quad \text{for all } A \in \mathcal{F}.$$

Unfortunately, the latter assumption (continuity of the group action) is missing in my former publications, which opens the door for pathologies.[23]

Remark 3.28. Continuity of the factorization follows from other assumptions, see [15, Lemma 2.1]. For arbitrary factorizations, continuity is restrictive (recall Example 3.19); waiving it, we get discontinuity points $t \in \mathbb{R}$ which are a finite or countable set. For a noise, however, the set is invariant under time shifts, and therefore, empty.

Lemma 3.29. *For every dyadic coarse factorization satisfying (3.13), its refinement is a noise.*

Proof. Our first argument parallels the proof of Lemma 3.11. Namely, let s, t be two coarse instants such that $s[\infty] = t[\infty]$. We introduce a coarse event r:

$$r[i] = \begin{cases} s[i] & \text{for } i \text{ even,} \\ t[i] & \text{for } i \text{ odd.} \end{cases}$$

We have

$$\text{Lim}\,\alpha_s[i] = \text{Lim}\,\alpha_s[2i] = \text{Lim}\,\alpha_r[2i] = \text{Lim}\,\alpha_r[i].$$

---

[23] Most results of these former publications do not depend on the (missing) continuity condition. But anyway, a discontinuous group action is a pathology, no doubt. (In particular, it cannot be Borel measurable.) The proof of Lemma 2.9 of [15], based on Weyl's relation, depends on the continuity condition.

Similarly, $\mathrm{Lim}\,\alpha_t[i] = \mathrm{Lim}\,\alpha_r[i]$. Thus, $\mathrm{Lim}\,\alpha_s[i] = \mathrm{Lim}\,\alpha_t[i]$, and we may define a one-parameter group of automorphisms $(\alpha_t)_{t\in\mathbb{R}}$ on $(\Omega, \mathcal{F}, P)$ by $\alpha_{t[\infty]} = \mathrm{Lim}\,\alpha_t[i]$.

Our second argument resembles the proof of Lemma 3.13. Namely, assume existence of $A_\infty \in \mathcal{F}$, $\varepsilon > 0$ and $t_n \to 0$ such that $P\big(A_\infty \,\triangle\, \alpha_{t_n}^{-1}(A_\infty)\big) \geq \varepsilon$ for all n. We choose a coarse event $A \in \mathcal{A}$ such that $A[\infty] = A_\infty$, and coarse instants s_n such that $s_n[\infty] = t_n$ for all n. Taking into account that $P[i]\big(A[i]\triangle\alpha_{s_n}^{-1}[i]A[i]\big) \to P\big(A_\infty\triangle\alpha_{t_n}^{-1}(A_\infty)\big) \geq \varepsilon$ and $s_n[i] \to t_n$ when $i \to \infty$, we choose integers $i_1 < i_2 < \ldots$ such that $P[i]\big(A[i] \,\triangle\, \alpha_{s_n}^{-1}[i]A[i]\big) \geq \varepsilon/2$ and $|s_n[i]| \leq |t_n| + 1/n$ whenever $i \geq i_n$. We define a coarse instant r by $r[i] = s_n[i]$ whenever $i_n \leq i < i_{n+1}$. Clearly, $r[\infty] = 0$; therefore $\mathrm{Lim}\,\alpha_r^{-1}[i]A[i] = \alpha_0^{-1}A[\infty] = A[\infty]$, and $P[i]\big(A[i] \,\triangle\, \alpha_r^{-1}[i]A[i]\big) \to 0$, which is impossible: these probabilities exceed $\varepsilon/2$. The contradiction proves continuity of the group $(\alpha_t)_{t\in\mathbb{R}}$. $\qquad\square$

Question 3.30. Is every noise the refinement of some dyadic coarse factorization satisfying (3.13)? I do not know; I guess that the answer is negative. It would be interesting to find some special features of such refinements among all noises. It is also unclear what happens to the class of such refinements, if subsequences are permitted (like in Example 3.9).

4 Example: The Noise Made by a Poisson Snake

This section is based on a paper by J. Warren entitled "The noise made by a Poisson snake" [23].

4.1 Three Discrete Semigroups: Algebraic Definition

A discrete semigroup (with unit; non-commutative, in general) may be defined by generators and relations.

Two generators f_+, f_- with two relations $f_+ f_- = 1$, $f_- f_+ = 1$ generate a semigroup G_1^{discrete} that is in fact a group, just the cyclic group $\mathbb{Z}$. Indeed, every word reduces to some f_+^k or f_-^k (or 1).

Two generators f_+, f_- with a single relation $f_+ f_- = 1$ generate a semigroup G_2^{discrete}. Every word reduces to some $f_-^k f_+^l$. The composition is

$$(f_-^{k_1} f_+^{l_1})(f_-^{k_2} f_+^{l_2}) = f_-^k f_+^l, \qquad \begin{aligned} k &= k_1 + \max(0, k_2 - l_1), \\ l &= l_2 + \max(0, l_1 - k_2). \end{aligned} \tag{4.1}$$

The canonical homomorphism $G_2^{\text{discrete}} \to G_1^{\text{discrete}}$ maps f_+ to f_+, f_- to f_-, and $f_-^k f_+^l$ into f_-^{k-l} (if $k > l$), f_+^{l-k} (if $k < l$), or 1 (if $k = l$). Accordingly, the composition law (4.1) satisfies

$$l - k = (l_1 - k_1) + (l_2 - k_2).$$

There is a more convenient pair of parameters, $a = l - k$, $b = k$; that is,[24]

$$f_{a,b} = f_-^b f_+^{a+b} \quad \text{for } a, b \in \mathbb{Z},\ b \geq 0,\ a + b \geq 0;$$

$$f_{a_1,b_1} f_{a_2,b_2} = f_{a,b}, \qquad \begin{aligned} a &= a_1 + a_2, \\ b &= \max(b_1, b_2 - a_1). \end{aligned} \tag{4.2}$$

The canonical homomorphism $G_2^{\text{discrete}} \to G_1^{\text{discrete}}$ maps $f_{a,b}$ to f_a, where $f_a \in G_1^{\text{discrete}}$ is f_+^a for $a > 0$, $f_-^{|a|}$ for $a < 0$, and 1 for $a = 0$.

Three generators f_-, f_+, f_* with three relations

$$f_+ f_- = 1, \quad f_* f_- = 1, \quad f_* f_+ = f_* f_* \tag{4.3}$$

generate a semigroup G_3^{discrete}. Every word reduces to some $f_-^k f_+^l f_*^m$. The following homomorphism $G_3^{\text{discrete}} \to G_2^{\text{discrete}}$ will be called canonical: $f_- \mapsto f_-$, $f_+ \mapsto f_+$, $f_* \mapsto f_+$. We have $f_-^k f_+^l f_*^m \mapsto f_-^k f_+^{l+m}$, which suggests such a triple of parameters for G_3^{discrete}: $a = l + m - k$, $b = k$, $c = m$; that is,

$$f_{a,b,c} = f_-^b f_+^{a+b-c} f_*^c \quad \text{for } a, b, c \in \mathbb{Z},\ b \geq 0,\ 0 \leq c \leq a + b;$$

$$f_{a_1,b_1,c_1} f_{a_2,b_2,c_2} = f_{a,b,c}, \qquad \begin{aligned} a &= a_1 + a_2, \\ b &= \max(b_1, b_2 - a_1), \end{aligned} \qquad c = \begin{cases} a_2 + c_1 & \text{if } c_1 > b_2, \\ c_2 & \text{otherwise.} \end{cases} \tag{4.4}$$

The canonical homomorphism $G_3^{\text{discrete}} \to G_2^{\text{discrete}}$ is just $f_{a,b,c} \mapsto f_{a,b}$.

Note that G_1^{discrete} is commutative, but G_2^{discrete} and G_3^{discrete} are not.

---

[24] Parameters a, b of (4.2) and a, b, c of (4.4) are suggested by S. Watanabe.

4.2 The Three Discrete Semigroups: Representation

By a representation of a semigroup G on a set S we mean a map $G \times S \ni (g, s) \mapsto g(s) \in S$ such that

$$(g_1 g_2)(s) = g_2(g_1(s)) \quad \text{and} \quad 1(s) = s$$

for all $g_1, g_2 \in G$, $s \in S$. The representation is called faithful, if

$$g_1 \neq g_2 \implies \exists s \in S \; (g_1(s) \neq g_2(s)).$$

Every G has a faithful representation on itself, $S = G$, namely, the regular representation, $g(g_0) = g_0 g$. Fortunately, G_2^{discrete} and G_3^{discrete} have more economical faithful representations on the set $\mathbb{Z}_+ = \{0, 1, 2, \dots\}$. Namely, for G_2^{discrete},

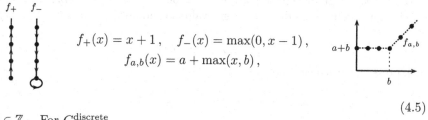

$$f_+(x) = x + 1, \quad f_-(x) = \max(0, x - 1),$$
$$f_{a,b}(x) = a + \max(x, b),$$

(4.5)

$x \in \mathbb{Z}_+$. For G_3^{discrete},

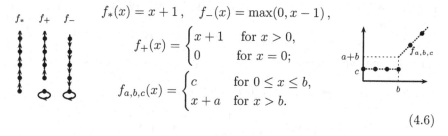

$$f_*(x) = x + 1, \quad f_-(x) = \max(0, x - 1),$$

$$f_+(x) = \begin{cases} x + 1 & \text{for } x > 0, \\ 0 & \text{for } x = 0; \end{cases}$$

$$f_{a,b,c}(x) = \begin{cases} c & \text{for } 0 \leq x \leq b, \\ x + a & \text{for } x > b. \end{cases}$$

(4.6)

4.3 Random Walks and Stochastic Flows in Discrete Semigroups

Example 4.1. The standard random walk on $\mathbb{Z}$ may be described by G_1^{discrete}-valued random variables

$$\xi_{s,t} = \xi_{s,s+1} \xi_{s+1,s+2} \cdots \xi_{t-1,t} \quad \text{for } s, t \in \mathbb{Z}, \; s \leq t;$$
$$\xi_{t,t+1} \text{ are independent random variables } (t \in \mathbb{Z});$$
$$\mathbb{P}\left(\xi_{t,t+1} = f_-\right) = \frac{1}{2} = \mathbb{P}\left(\xi_{t,t+1} = f_+\right) \quad \text{for each } t \in \mathbb{Z}.$$

(4.7)

Note that $\xi_{r,s} \xi_{s,t} = \xi_{r,t}$ whenever $r \leq s \leq t$. Everyone knows that

$$\mathbb{P}\left(\xi_{0,t} = f_a\right) = \frac{1}{2^t} \binom{t}{\frac{t+a}{2}}$$

(4.8)

for $a = -t, -t+2, -t+4, \ldots, t$.

In fact, 'the standard random walk' is the random process $t \mapsto \xi_{0,t}$. Taking into account that G_1^{discrete} is a group, $\xi_{s,t}$ may be thought of as an increment, $\xi_{s,t} = \xi_{0,s}^{-1}\xi_{s,t}$.

Example 4.2. Formulas (4.7) work equally well on G_2^{discrete}. Still, $\xi_{r,s}\xi_{s,t} = \xi_{r,t}$. However, G_2^{discrete} is not a group, and $\xi_{s,t}$ is not an increment; moreover, it is not a function of $\xi_{0,s}$ and $\xi_{0,t}$. Indeed, knowing a_1, b_1 and $a_1 + a_2$, $\max(b_1, b_2 - a_1)$ (recall (4.2)) we can find a_2 but not b_2. Thus, the two-parameter family $(\xi_{s,t})_{s \le t}$ of random variables is more than just a random walk. Let us call such a family an *abstract stochastic flow*. Why 'abstract'? Since G_2^{discrete} is an abstract semigroup rather than a semigroup of transformations (of some set). So, we have the standard abstract flow in G_2^{discrete}. In order to get a (usual, not abstract) stochastic flow, we have to choose a representation of G_2^{discrete}. Of course, the regular representation could be used, but the representation (4.5) is more useful. Introducing integer-valued random variables $a(s,t), b(s,t)$ by

$$\xi_{s,t} = f_{a(s,t),b(s,t)}$$

we express the stochastic flow as

$$\xi_{s,t}(x) = a(s,t) + \max(x, b(s,t)).$$

Fixing s and x we get a random process called a single-point motion of the flow. Namely, it is a reflecting random walk. Especially, for $s = 0$ and $x = 0$, the process

$$t \mapsto \xi_{0,t}(0) = a(0,t) + b(0,t)$$

is a reflecting random walk. It is easy to see that two processes

$$t \mapsto \xi_{0,t}(0) = a(0,t) + b(0,t),$$

$$t \mapsto \left| a(0,t) + \frac{1}{2} \right| - \frac{1}{2}$$

are identically distributed. Also,

$$b(0,t) = - \min_{s=0,1,\ldots,t} a(0,s),$$

$$a(0,t) + b(0,t) = \max_{s=0,1,\ldots,t} a(s,t),$$

$$\tag{4.9}$$

and $a(\cdot, \cdot)$ is the standard random walk on $G_1^{\text{discrete}} = \mathbb{Z}$. That is, the canonical homomorphism $G_2^{\text{discrete}} \to G_1^{\text{discrete}}$ transforms the standard flow on G_2^{discrete} into the standard flow (or random walk) on G_1^{discrete}. Using the reflection principle, one gets

$$\mathbb{P}\left(\xi_{0,t} = f_{a,b}\right) = \frac{a + 2b + 1}{2^t} \cdot \frac{t!}{\left(\frac{t+a}{2} + b + 1\right)!\left(\frac{t-a}{2} - b\right)!}. \qquad (4.10)$$

Note that a, b occur only in the combination $a + 2b$.

Example 4.3. On G_3^{discrete}, we have no 'standard' random walk or flow; rather, we introduce a one-parameter family of abstract stochastic flows,

$$\xi_{s,t} = \xi_{s,s+1}\xi_{s+1,s+2}\cdots\xi_{t-1,t} \quad \text{for } s, t \in \mathbb{Z}, \; s \le t;$$

$$\xi_{t,t+1} \text{ are independent random variables } (t \in \mathbb{Z});$$

$$\mathbb{P}\left(\xi_{t,t+1} = f_-\right) = \frac{1}{2}, \quad \mathbb{P}\left(\xi_{t,t+1} = f_+\right) = \frac{1-p}{2}, \quad \mathbb{P}\left(\xi_{t,t+1} = f_*\right) = \frac{p}{2};$$
$$(4.11)$$

$p \in (0, 1)$ is the parameter. The canonical homomorphism $G_3^{\text{discrete}} \to G_2^{\text{discrete}}$ glues together f_+ and f_*, thus eliminating the parameter p and giving the standard abstract flow on G_2^{discrete}. Defining $a(\cdot, \cdot), b(\cdot, \cdot), c(\cdot, \cdot)$ by

$$\xi_{s,t} = f_{a(s,t),b(s,t),c(s,t)}$$

we see that the joint distribution of $a(\cdot, \cdot)$ and $b(\cdot, \cdot)$ is the same as before.

Representation (4.6) of G_3^{discrete} turns the abstract flow into a stochastic flow on $\mathbb{Z}_+$. Its single-point motion is a sticky random walk,

$$t \mapsto \xi_{0,t}(0) = c(0, t).$$

In order to find the conditional distribution of $c(\cdot, \cdot)$ given $a(\cdot, \cdot)$ and $b(\cdot, \cdot)$ we observe that

$$a(0, t) - c(0, t) = \min\left(a(0, t), \min\{x : \xi_{\sigma(x),\sigma(x)+1} = f_*\}\right) \qquad (4.12)$$
$$\text{where } \sigma(x) = \max\{s = 0, \ldots, t : a(0, s) = x\}, \quad -b(0, t) \le x < a(0, t).$$

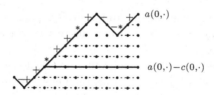

Therefore the conditional distribution of $c(0, t)$ is basically the truncated geometric distribution. More exactly, it is the (conditional) distribution of

$$\max\left(0, a(0, t) + b(0, t) - G + 1\right), \qquad G \sim \text{Geom}(p); \qquad (4.13)$$

here G is a random variable, independent of $a(\cdot, \cdot), b(\cdot, \cdot)$, such that $\mathbb{P}\left(G = g\right) = p(1-p)^{g-1}$ for $g = 1, 2, \ldots$ This is the discrete counterpart of a well-known result of J. Warren [21]. So,

$$\mathbb{P}\left(\xi_{0,t} = f_{a,b,c}\right) = \frac{a + 2b + 1}{2^t} \frac{t!}{\left(\frac{t+a}{2} + b + 1\right)!\left(\frac{t-a}{2} - b\right)!} \cdot p(1-p)^{a+b-c}$$

(4.14)

for $c > 0$; for $c = 0$ the factor $p(1-p)^{a+b-c}$ turns into $(1-p)^{a+b}$, rather than $p(1-p)^{a+b}$, because of truncation.

4.4 Three Continuous Semigroups

The continuous counterpart of the discrete semigroup $G_1^{\text{discrete}} = \mathbb{Z}$ is the semigroup $G_1 = \mathbb{R} = \{f_a : a \in \mathbb{R}\}$, $f_{a_1} f_{a_2} = f_{a_1+a_2}$.

The continuous counterpart of the discrete semigroup $G_2^{\text{discrete}} = \{f_{a,b} : a, b \in \mathbb{Z}, b \geq 0, a + b \geq 0\}$ is the semigroup

$$G_2 = \{f_{a,b} : a, b \in \mathbb{R}, b \geq 0, a + b \geq 0\},$$

$$f_{a_1,b_1} f_{a_2,b_2} = f_{a,b}, \quad \begin{aligned} a &= a_1 + a_2, \\ b &= \max(b_1, b_2 - a_1) \end{aligned}$$

(4.15)

(recall (4.2)). The canonical homomorphism $G_2 \to G_1$ maps $f_{a,b}$ to f_a.

The continuous counterpart of the discrete semigroup $G_3^{\text{discrete}} = \{f_{a,b,c} : a, b, c \in \mathbb{Z}, b \geq 0, 0 \leq c \leq a + b\}$ is the semigroup

$$G_3 = \{f_{a,b,c} : a, b, c \in \mathbb{R}, b \geq 0, 0 \leq c \leq a + b\},$$

$$f_{a_1,b_1,c_1} f_{a_2,b_2,c_2} = f_{a,b,c}, \quad \begin{aligned} a &= a_1 + a_2, \\ b &= \max(b_1, b_2 - a_1), \end{aligned} \quad c = \begin{cases} a_2 + c_1 & \text{if } c_1 > b_2, \\ c_2 & \text{otherwise} \end{cases}$$

(4.16)

(recall (4.4)). The canonical homomorphism $G_3 \to G_2$ maps $f_{a,b,c}$ to $f_{a,b}$.

Note that G_1 is commutative but G_2, G_3 are not. Also, G_1 and G_2 are topological semigroups, but G_3 is not (since the composition is discontinuous at $c_1 = b_2$).

There are two one-parameter semigroups in G_2, $\{f_{a,0} : a \in [0, \infty)\}$ and $\{f_{-b,b} : b \in [0, \infty)\}$. They generate G_2 according to the relation $f_{b,0} f_{-b,b} = 1$; namely, $f_{a,b} = f_{-b,b} f_{a+b,0}$.

There are three one-parameter semigroups in G_3, $\{f_{a,0,0} : a \in [0, \infty)\}$, $\{f_{-b,b,0} : b \in [0, \infty)\}$ and $\{f_{c,0,c} : c \in [0, \infty)\}$. They generate G_3 according to relations $f_{b,0,0} f_{-b,b,0} = 1$, $f_{b,0,0} f_{-b,b,0} = 1$, and $f_{c,0,c} f_{a,0,0} = f_{c,0,c} f_{a,0,a}$ for $c > 0$; namely, $f_{a,b,c} = f_{-b,b,0} f_{a+b-c,0,0} f_{c,0,c}$.

Here is a faithful representation of G_2 on $[0, \infty)$ (recall (4.5)):

$$f_{a,b}(x) = a + \max(x, b),$$

(4.17)

$x \in [0, \infty)$.

Here is a faithful representation of G_3 on $[0, \infty)$ (recall (4.6)):

$$f_{a,b,c}(x) = \begin{cases} c & \text{for } 0 \le x \le b, \\ x + a & \text{for } x > b. \end{cases} \tag{4.18}$$

All functions are increasing, but $f_{a,b}$ are continuous, while $f_{a,b,c}$ are not.

4.5 Convolution Semigroups in These Continuous Semigroups

Example 4.4. Everyone knows that the binomial distribution (4.8) is asymptotically normal. That is, the distribution of $\sqrt{\varepsilon}a(0, t/\varepsilon)$ converges weakly (for $\varepsilon \to 0$) to the normal distribution $\mu_t^{(1)} = N(0, t)$. These form a convolution semigroup, $\mu_s^{(1)} * \mu_t^{(1)} = \mu_{s+t}^{(1)}$.

Note however, that $a(s, t)$ and $\xi_{s,t}$ are defined (see (4.7)) only for integers s, t. We may extend them, in one way or another, to real s, t. Or alternatively, we may use coarse instants $t = (t[i])_{i=1}^{\infty}$, $t[i] \in \frac{1}{i}\mathbb{Z}$, $t[i] \to t[\infty]$, introduced in Sect. 3.2. For every coarse instant t, the distribution of $i^{-1/2}a(0, it[i])$ converges weakly (for $i \to \infty$) to $\mu_{t[\infty]}^{(1)} = N(0, t[\infty])$.

Example 4.5. The two-dimensional distribution (4.10) on G_2^{discrete} has its asymptotics. Namely, the joint distribution of $i^{-1/2}a(0, it[i])$ and $i^{-1/2}b(0, it[i])$ converges weakly (for $i \to \infty$) to the measure $\mu_{t[\infty]}^{(2)}$ with density (on the relevant domain $b > 0$, $a + b > 0$; t means $t[\infty]$):

$$\frac{\mu_t^{(2)}(dadb)}{dadb} = \frac{2(a + 2b)}{\sqrt{2\pi}\,t^{3/2}} \exp\left(-\frac{(a + 2b)^2}{2t}\right). \tag{4.19}$$

Treating $\mu_t^{(2)}$ (for $t \in [0, \infty)$) as a measure on G_2, we get a convolution semigroup: $\mu_s^{(2)} * \mu_t^{(2)} = \mu_{s+t}^{(2)}$. Of course, the convolution is taken according to the composition (4.15).

Example 4.6. What about the three-dimensional distribution (4.14) on G_3^{discrete}? It has a parameter p. In order to get a non-degenerate asymptotics, we let p depend on i, namely,

$$p = \frac{1}{\sqrt{i}} \to 0.$$

Then the distribution of $i^{-1/2}G$, where $G \sim \text{Geom}(p)$ (recall (4.13)), converges weakly to the exponential distribution $\text{Exp}(1)$, and the joint distribution of $i^{-1/2}a(0, it[i])$, $i^{-1/2}b(0, it[i])$ and $i^{-1/2}c(0, it[i])$ converges weakly to a measure $\mu_{t[\infty]}^{(3)}$. The measure has an absolutely continuous part and a singular part (at $c = 0$), and may be described (somewhat indirectly) as the joint distribution of three random variables a, b and $(a + b - \eta)^+$, where the pair (a, b) is distributed $\mu_t^{(2)}$ (see (4.19)), η is independent of (a,b), and $\eta \sim \mathrm{Exp}(1)$. Treating $\mu_t^{(3)}$ (for $t \in [0, \infty)$) as a measure on G_3, we get a convolution semigroup: $\mu_s^{(3)} * \mu_t^{(3)} = \mu_{s+t}^{(3)}$, the convolution being taken according to the composition (4.16). No need to check the relation 'by hand'; it follows from its discrete counterpart. The latter follows from the construction of Sect. 4.3 (since random variables $\xi_{0,1}, \xi_{1,2}, \ldots, \xi_{s+t-1,s+t}$ are independent). It may seem that the limiting procedure does not work, since G_3 is not a topological semigroup; the composition (4.16) is discontinuous at $c_1 = b_2$. However, that is not an obstacle, since the equality $c_1 = b_2$ is of zero probability, as far as triples (a_1, b_1, c_1) and (a_2, b_2, c_2) are independent and distributed $\mu_s^{(3)}$, $\mu_t^{(3)}$, respectively $(s, t > 0)$. The atom of c_1 at 0 does not matter, since b_2 is nonatomic. The composition is continuous almost everywhere!

4.6 Getting Dyadic

Our flows in G_1^{discrete} and G_2^{discrete} are dyadic (two equiprobable possibilities in each step), which cannot be said about G_3^{discrete}; here, in each step, we have three possibilities f_-, f_+, f_* of probabilities $1/2, (1-p)/2, p/2$. Can a dyadic model produce the same asymptotic behavior? Yes, it can, at the expense of using $i \in \{1, 4, 16, 64, \ldots\}$ only (recall Example 3.9); and, of course, the dyadic model is more complicated.[25] Instead of the trap at 0, we design a trap near 0 as follows:

$$g_+ = f_* = f_{1,0,1}; \quad g_- = f_-^m f_+^{m-1} = f_{-1,m,0};$$
$$\mathbb{P}\left(\xi_{t,t+1} = g_-\right) = \frac{1}{2} = \mathbb{P}\left(\xi_{t,t+1} = g_+\right).$$

The old (small) parameter p disappears, and a new (large) parameter m appears. We'll see that the two models are asymptotically equivalent, when $p = 2^{-m}$.

As before, we may denote

$$\xi_{s,t} = f_{a(s,t),b(s,t),c(s,t)}.$$

Note, however, that only $a(s,t)$ is the same as before; $b(s,t)$, $c(s,t)$ and $\xi_{s,t}$ are modified. Formula (4.9) for $b(0,t)$ fails, but still,

$$b(0,t) = -\min_{s=0,1,\ldots,t} a(0,s) + O(m), \tag{4.20}$$

which is asymptotically the same. Formula (4.12) for $c(0,t)$ also fails. Instead,

---

[25] Maybe, a still more complicated construction can use all i; I do not know.

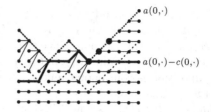

$$a(0,t) - c(0,t) = \min\{x : \sigma(x + m - 1) - \sigma(x) = m - 1\}, \qquad (4.21)$$

if such x exists in the set $\mathbb{Z} \cap [\min_{[0,t]} a(0,\cdot), a(0,t) - m + 1]$; otherwise, $c(0,t) = O(m)$. (Here σ is the same as in (4.12).)

The conditional distribution of $c(0,t)$, given the path $a(0,\cdot)$, is not at all geometric (unlike (4.13)), since now $c(0,t)$ is uniquely determined by $a(0,\cdot)$. However, according to (4.21), $a(0,t) - c(0,t)$ is determined by small increments of the process $\sigma(\cdot)$. On the other hand, the large-scale structure of the path $a(0,\cdot)$ is correlated mostly with large increments of $\sigma(\cdot)$; small increments are numerous, but contribute little to the sum. Using this argument, one can show that $c(0,t)$ is asymptotically independent of $a(0,t)$ (and $b(0,t)$, due to (4.20)).

The unconditional distribution of $c(0,t)$ can be found from (4.21), taking into account that increments $\sigma(x + 1) - \sigma(x)$ are independent, and each increment is equal to 1 with probability $1/2$. We have Bernoulli trials, and we wait for the first block of $m - 1$ 'successes'. For large m, the waiting time is approximately exponential, with the mean 2^m.[26] Thus, $2^{-m}(a(0,t) - c(0,t) - \min_{[0,t]} a(0,\cdot))$ is asymptotically $\mathrm{Exp}(1)$, truncated (at $c = 0$) as in Sect. 4.5.

Taking the limit $i = 2^{2m} \to \infty$, we get for $i^{-1/2} a(0, it[i])$, $i^{-1/2} b(0, it[i])$, $i^{-1/2} c(0, it[i])$ the limiting distribution $\mu^{(3)}_{t[\infty]}$, the same as in Sect. 4.5.

4.7 Scaling Limit

For any coarse instants s, t such that $s \leq t$, the distribution $\mu^{(n)}_{s,t}[i]$ of $i^{-1/2} \xi^{(n)}_{is[i], it[i]}$ converges weakly (for $i \to \infty$) to the measure $\mu^{(n)}_{s,t}[\infty] = \mu^{(n)}_{t[\infty] - s[\infty]}$ on G_n, for our three models, $n = 1, 2, 3$. Of course, multiplication of ξ by $i^{-1/2}$ is understood as multiplication of $a(\cdot, \cdot)$, $b(\cdot, \cdot)$, $c(\cdot, \cdot)$ by $i^{-1/2}$, which is a homomorphic embedding of G_n^{discrete} into G_n.

Let r, s, t be coarse instants, $r \leq s \leq t$. Due to independence, the joint distribution $\mu^{(n)}_{r,s}[i] \otimes \mu^{(n)}_{s,t}[i]$ of random variables $i^{-1/2} \xi^{(n)}_{ir[i], is[i]}$ and $i^{-1/2} \xi^{(n)}_{is[i], it[i]}$ converges weakly to $\mu^{(n)}_{r,s}[\infty] \otimes \mu^{(n)}_{s,t}[\infty]$. However, we need the joint distribution of three random variables,

$$i^{-1/2} \xi^{(n)}_{ir[i], is[i]}, \quad i^{-1/2} \xi^{(n)}_{is[i], it[i]}, \quad i^{-1/2} \xi^{(n)}_{ir[i], it[i]},$$

---

[26] Such a block appears, in the mean, after 2^{m-1} shorter blocks, of mean length ≈ 2 each.

the third being the product of the first and the second in the semigroup G_n. For $n = 1, 2$ weak convergence for the triple follows immediately from weak convergence for the pair, since the composition is continuous. For $n = 3$, discontinuity of the composition in G_3 does not invalidate the argument, since the composition is continuous almost everywhere w.r.t. the relevant measure (recall Sect. 4.5).

Similarly, for every k and all coarse instants $t_1 \leq \cdots \leq t_k$, the joint distribution of $k(k-1)/2$ random variables $i^{-1/2}\xi^{(n)}_{it_l[i],it_m[i]}$, $1 \leq l < m \leq k$, converges weakly (for $i \to \infty$). We choose a sequence $(t_k)_{k=1}^{\infty}$ of coarse instants such that the sequence of numbers $(t_k[\infty])_{k=1}^{\infty}$ is dense in $\mathbb{R}$, and use Lemma 2.17, getting a coarse probability space.

The Hölder condition, the same as in Example 2.1, holds for all three models. I mean Hölder continuity of $a(\cdot, \cdot)$, $b(\cdot, \cdot)$, $c(\cdot, \cdot)$. Indeed, $a(\cdot, \cdot)$ is the same as in Example 2.1; $b(\cdot, \cdot)$ is related to $a(\cdot, \cdot)$ via (4.9) or (4.20), and $c(\cdot, \cdot)$ satisfies (on any interval)

$$\max_{|s-t| \leq x} |c(0,s) - c(0,t)| \leq \max_{|s-t| \leq x} |a(0,s) - a(0,t)|,$$

though, for the model of Sect. 4.6, $O(m)$ must be added.

Thus, a joint σ-compactification is constructed for all three models (the third model — in two versions, Example (4.3) and Sect. 4.6).

4.8 Noises

Example 4.7. The standard flow in G_1^{discrete}, rescaled by $i^{-1/2}$, gives us a coarse probability space, identical to that of Example 3.6. It is a dyadic coarse factorization. Its refinement is the Brownian continuous factorization. Equipped with the natural time shift, it is a noise.

Example 4.8. The standard flow in G_2^{discrete}, rescaled by $i^{-1/2}$, gives us another coarse probability space. It is also a dyadic coarse factorization (the proof is similar to the previous case). Its 'two-dimensional nature' is a delusion; the dyadic coarse factorization is identical to that of Example 4.7. The second dimension $b(\cdot, \cdot)$ reduces to the first dimension, $a(\cdot, \cdot)$, by (4.9).

Example 4.9. The flow in G_3, introduced in Example 4.3, rescaled by $i^{-1/2}$ with $p = i^{-1/2}$ (recall Example 4.6), gives us a coarse probability space. It is not a dyadic coarse factorization, since it is not dyadic. However, it satisfies a natural generalization of Definition 3.4 to the non-dyadic case (the proof is as before). Its refinement is a continuous factorization, and (with natural time shift), a noise; it may be called the noise of stickiness.

Once again, the second dimension, $b(\cdot, \cdot)$, reduces to the first dimension, $a(\cdot, \cdot)$. Indeed, the joint distribution of $a(\cdot, \cdot)$ and $b(\cdot, \cdot)$ is the same as in Example 4.8. What about the third dimension, $c(\cdot, \cdot)$?

The conditional distribution of $c(s,t)$, given $a(s,t)$ and $b(s,t)$, is basically truncated exponential. Namely, it is the distribution of $\big(a(s,t) + b(s,t) - \eta\big)^+$

where $\eta \sim \text{Exp}(1)$; see Example 4.6. Moreover, for any $r < s < t$, the conditional distribution of $c(r,t)$ given $a(r,s), b(r,s)$ and $a(s,t), b(s,t)$, is still the distribution of $\big(a(r,t) + b(r,t) - \eta\big)^+$. In other words, $c(r,t)$ is conditionally independent of $a(r,s), b(r,s), a(s,t), b(s,t)$, given $a(r,t), b(r,t)$. That is a property of the composition (4.16); if $c_1 \sim \big(a_1 + b_1 - \eta_1\big)^+$ and $c_2 \sim \big(a_2 + b_2 - \eta_2\big)^+$ then $c \sim \big(a + b - \eta\big)^+$.

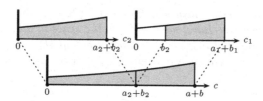

It follows by induction that the conditional distribution of $c(t_1, t_n)$, given all $a(t_i, t_j)$ and $b(t_i, t_j)$, is given by the same formula $\big(a(t_1, t_n) + b(t_1, t_n) - \eta\big)^+$, $\eta \sim \text{Exp}(1)$, for every n and $t_1 < \cdots < t_n$. Therefore, the same holds for the conditional distribution of $c(s,t)$ given all $a(u,v)$ and $b(u,v)$ for u, v such that $s \le u \le v \le t$ (a well-known result of J. Warren [21]). We see that $c(\cdot, \cdot)$ is not a function of $a(\cdot, \cdot)$ (and $b(\cdot, \cdot)$).

Example 4.10. Another flow in G_3^{discrete}, introduced in Sect. 4.6, being rescaled by $i^{-1/2}$ with $i = 2^{2m}$, gives us a dyadic coarse factorization. Its refinement is the same continuous factorization (and noise) as in Example 4.9.

4.9 The Poisson Snake

Formula (4.12) suggests a description of the sticky flow in G_3^{discrete} by a combination of a simple random walk $a(\cdot, \cdot)$ and a random subset of the set of its 'chords'. A chord may be defined as an interval $[s,t]$, $s, t \in \mathbb{Z}$, $s < t$, such that $a(s,t) = 0$ and $a(s,u) > 0$ for all $u \in (s,t) \cap \mathbb{Z}$. Or equivalently, a chord is a horizontal straight segment on the plane that connects points $\big(s, a(0,s)\big)$ and $\big(t, a(0,t)\big)$ and goes below the graph of $a(0, \cdot)$. The random subset of chords is very simple: every chord belongs to the subset with probability p, independently of others. Note that $p = i^{-1/2}$ is equal to the vertical pitch (after rescaling $a(\cdot, \cdot)$ by $i^{-1/2}$). The scaling limit suggests itself: a Poisson random subset of the set of all chords of the Brownian sample path.

Definition 4.11. *A finite chord of a continuous function $f : \mathbb{R} \to \mathbb{R}$ is a set of the form $[s,t] \times \{x\} \subset \mathbb{R}^2$ where $s < t$, $x = f(s)$ and $t = \inf\{u \in (s, \infty) : f(u) \le x\}$. An infinite chord of f is a set of the form $[s, \infty) \times \{x\} \subset \mathbb{R}^2$ where $x = f(s)$ and $f(t) > x$ for all $t \in (s, \infty)$. A chord of f is either a finite chord of f, or an infinite chord of f.*

a chord   a chord

If f decreases, it has no chords. Otherwise it has a continuum of chords. The set of chords is, naturally, a standard Borel space,[27] due to the one-one correspondence between a chord and its initial point $(s, x) \in \mathbb{R}^2$.

Lemma 4.12. *For every continuous function $f : \mathbb{R} \to \mathbb{R}$ there exists one and only one σ-finite positive Borel measure[28] on the space of all chords of f, such that the set of chords that intersect a vertical segment $\{t\} \times [x, y]$ is of measure $y - x$, whenever t, x, y are such that $\inf_{s \in (-\infty, t)} f(s) \le x < y \le f(t)$.*

The proof is left to the reader. Hint: for every $\varepsilon > 0$, the set of chords longer than ε is elementary; on this set, the measure is locally finite.

The map $[s, t] \times \{x\} \mapsto s$ (also $[s, \infty) \times \{x\} \mapsto s$, of course) sends the measure on the set of chords (described in Lemma 4.12) into a measure on $\mathbb{R}$. If f is of locally finite variation, then the measure on $\mathbb{R}$ is just $(df)^+$, the positive part of the Lebesgue-Stieltjes measure. However, we need the opposite case: f is of infinite variation on every interval, and the measure is also infinite on every interval. Nevertheless, it is σ-finite (but not locally finite). We denote it $(df)^+$ anyway.

The measure $(df)^+$ is concentrated on the set of points of 'local minimum from the right'. If f is a Brownian sample path then such points are a set of Lebesgue measure 0.

So, the set of all chords is a measure space; it carries a natural σ-finite (sometimes, finite) measure. The latter is the intensity measure of a unique Poisson random measure.[29] This way, (the distribution of) a random set of chords is well-defined.

Or equivalently, we may consider a Poisson random subset of $\mathbb{R}$, whose intensity measure is $(df)^+$.

However, it is not so easy to substitute a Brownian sample path $B(\cdot)$ for $f(\cdot)$. In order to get a (Poisson) random variable, we may ask how many random points belong to a given Borel set $A \subset \mathbb{R}$ such that $(dB)^+(A) < \infty$. Note that for any *interval* A, $(dB)^+(A) = \infty$ a.s. We cannot choose an appropriate A without knowing the path $B(\cdot)$. The set of all countable dense subsets of $\mathbb{R}$ does not carry a natural (non-pathological) Borel structure.

---

[27] For a definition, see [7, Sect. 12.B].

[28] For a definition, see [7, Sect. 17.A].

[29] See for instance [11, XII.1.18].

In this aspect, chords are better than points. Chords are parameterized by three (or two) numbers, and thus, carry a natural Borel structure, irrespective of $B(\cdot)$. The random countable set of chords is not dense; rather, it accumulates toward short chords.

A point (t, x) belongs to a random chord of $B(\cdot)$ if and only if

$$x \in \sigma_t^{-1}(\Pi), \quad \text{that is,} \quad \sigma_t(x) \in \Pi,$$

$$\text{where } \sigma_t(x) = \sup\{s \in (-\infty, t] : B(s) \le x\} \text{ for } x \in (-\infty, B(t))$$

(recall (4.12)), and Π is the Poisson random subset of $\mathbb{R}$, whose intensity measure is $(dB)^+$. Do not confuse the inverse image $\sigma_t^{-1}(\Pi)$ with the image $B(\Pi)$. True, $B(\sigma_t(x)) = x$, but $\sigma_t(B(s)) \ne s$. Sets Π and $B(\Pi)$ are dense, but the set $\sigma_t^{-1}(\Pi)$ is locally finite. Moreover, $\sigma_t^{-1}(\Pi)$ is a Poisson random subset of $(-\infty, B(t)]$, its intensity being just 1.

The random countable dense set Π itself is bad; we have no measurable functions of it. However, the pair $(B(\cdot), \Pi)$ of the Brownian path and the set is good; we have measurable functions of the pair. In particular, we may use measurable functions of the locally finite set $\sigma_t^{-1}(\Pi)$. Especially,

$$a(0, t) - c(0, t) = \min\left(a(0, t), \min\{x : \sigma_t(x) \in \Pi \cap (0, \infty)\}\right).$$

Lemma 4.13. *The σ-field $\mathcal{F}_{s,t}$ of the noise of stickiness (see Example 4.9) is generated by Brownian increments $B(u) - B(s)$ for $u \in (s, t)$ and random sets $\sigma_u^{-1}(\Pi \cap (s, t))$ for $u \in (s, t)$ (treated as random variables whose values are finite subsets of $\mathbb{R}$).*

The proof is left to the reader.

5 Stability

5.1 Discrete Case

Fourier-Walsh coefficients, introduced in Sect. 3.3 for an arbitrary dyadic coarse factorization,

$$f = \sum_{M \in \mathcal{C}[i]} \hat{f}_M \tau_M = \hat{f}_\emptyset + \sum_{m \in \frac{1}{i}\mathbb{Z}} \hat{f}_{\{m\}} \tau_m + \sum_{m_1, m_2 \in \frac{1}{i}\mathbb{Z}, m_1 < m_2} \hat{f}_{\{m_1, m_2\}} \tau_{m_1} \tau_{m_2} + \dots$$

help us to examine the stability of a function f, as explained below. Imagine another array of random signs $(\tau'_m)_{m \in \frac{1}{i}\mathbb{Z}}$ (also independent equiprobable ± 1) correlated with the array $(\tau_m)_{m \in \frac{1}{i}\mathbb{Z}}$,

$$\mathbb{E}\, \tau_m \tau'_m = \rho \quad \text{for each } m \in \frac{1}{i}\mathbb{Z};$$

$\rho \in [-1, +1]$ is a parameter. Other correlations vanish. That is, the joint distribution of all τ_m and τ'_m is the product (over $m \in \frac{1}{i}\mathbb{Z}$) of (copies of) such a four-atom distribution:

$$
\begin{array}{c}
& \tau_m \\
& \begin{array}{cc} -1 & +1 \end{array} \\
\tau'_m \begin{array}{c} -1 \\ +1 \end{array} &
\begin{array}{|c|c|}
\hline
\frac{1+\rho}{4} & \frac{1-\rho}{4} \\
\hline
\frac{1-\rho}{4} & \frac{1+\rho}{4} \\
\hline
\end{array}
\end{array}
$$

Denoting by $\tilde{\Omega}[i]$ the product of these four-point probability spaces, we have a natural measure preserving map $\alpha : \tilde{\Omega}[i] \to \Omega[i]$; as before, $\Omega[i]$ is the product of two-point probability spaces. In addition, we have another measure preserving map $\alpha' : \tilde{\Omega}[i] \to \Omega[i]$,

$$\tau_m \circ \alpha = \tau_m, \quad \tau_m \circ \alpha' = \tau'_m;$$

we use the same 'τ_m' for denoting a coordinate function on $\Omega[i]$ and $\tilde{\Omega}[i]$.
For products

$$\tau_M = \prod_{m \in M} \tau_m, \quad M \in \mathcal{C}[i], \quad \mathcal{C}[i] = \{M \subset \tfrac{1}{i}\mathbb{Z} : |M| < \infty\}$$

we have

$$\mathbb{E}\, \tau_M \tau'_M = \rho^{|M|}, \quad \tau_M \circ \alpha = \tau_M, \quad \tau_M \circ \alpha' = \tau'_M,$$

where $|M|$ is the number of elements of M. Therefore

$$\mathbb{E}\, (f \circ \alpha)(g \circ \alpha') = \sum_M \rho^{|M|} \hat{f}_M \hat{g}_M = \langle g, \rho^{\mathbf{N}[i]} f \rangle,$$

$$\rho^{\mathbf{N}[i]} : L_2[i] \to L_2[i], \quad \rho^{\mathbf{N}[i]} \tau_M = \rho^{|M|} \tau_M, \quad \rho^{\mathbf{N}[i]} f = \sum_M \rho^{|M|} \hat{f}_M \tau_M.$$

The Hermite operator $\rho^{\mathbf{N}[i]}$ is a function of a self-adjoint operator $\mathbf{N}[i]$ defined by $\mathbf{N}[i]\tau_M = |M|\tau_M$ for $M \in \mathcal{C}[i]$.

Every bounded function $\varphi : \mathcal{C}[i] \to \mathbb{R}$ acts on $L_2[i]$ by the operator $f \mapsto \sum_{M \in \mathcal{C}[i]} \varphi(M) \hat{f}_M \tau_M$. A commutative operator algebra is isomorphic to the algebra of functions. The operator $\rho^{\mathbf{N}[i]}$ corresponds to the function $M \mapsto \rho^{|M|}$. (In some sense, the unbounded operator $\mathbf{N}$ corresponds to the unbounded function $M \mapsto |M|$.)

A function $\varphi : \mathcal{C}[i] \to \{0,1\}$, the indicator of a subset of $\mathcal{C}[i]$, corresponds to a projection operator. Say, for the (indicator of) the set $\{\emptyset\}$, the operator projects to the one-dimensional space of constants (the expectation). For the set $\{M : M \subset (0,\infty)\}$, the operator is the conditional expectation, $\mathbb{E}\left(\cdot \,\middle|\, \mathcal{F}_{0,\infty}[i] \right)$.

The function $M \mapsto |M|$ is the sum (over $m \in \frac{1}{i}\mathbb{Z}$) of localized functions $M \mapsto |M \cap \{m\}|$. The latter is the indicator of the set $\{M : M \ni m\}$, corresponding to the projection operator $1 - \mathbb{E}\left(\cdot \,\middle|\, \mathcal{F}_{\frac{1}{i}\mathbb{Z}\setminus\{m\}} \right)$. Thus,

$$\mathbf{N}f = \sum_m \left(f - \mathbb{E}\left(f \,\middle|\, \mathcal{F}_{\frac{1}{i}\mathbb{Z}\setminus\{m\}} \right) \right).$$

The operator $\rho^{\mathbf{N}[i]}$ may be interpreted as the conditional expectation w.r.t. the sub-σ-field $\alpha^{-1}(\mathcal{F})$ generated by $\tau_m \circ \alpha$, $m \in \frac{1}{i}\mathbb{Z}$:

$$\mathbb{E}\left(f \circ \alpha' \,\middle|\, \alpha^{-1}(\mathcal{F}) \right) = (\rho^{\mathbf{N}[i]} f) \circ \alpha \quad \text{for } f \in L_2[i].$$

We may imagine that our data τ_m are an unreliable copy of the true data τ'_m; each sign τ_m is either correct (with probability $(1+\rho)/2$) or inverted (with probability $(1-\rho)/2$). If ρ is close to 1, our knowledge of τ'_M is satisfactory for moderate $|M|$ (when $\rho^{|M|} \approx 1$) but very bad for large $|M|$ (when $\rho^{|M|} \approx 0$). The position of a given function f between the two extremes is indicated by the number $\|f - \rho^{\mathbf{N}} f\|$.

Example 5.1. In the Brownian coarse factorization (recall Example 3.6),

$$\sup_i \| f[i] - \rho^{\mathbf{N}[i]} f[i] \| \to 0 \quad \text{for } \rho \to 1$$

for all $f \in L_2(\mathcal{A})$. This follows easily from convergence of operators (recall Sect. 2.3 and Example 3.18):

$$\mathrm{Lim}_{i \to \infty} \rho^{\mathbf{N}[i]} = \rho^{\mathbf{N}[\infty]},$$

$$\rho^{\mathbf{N}[\infty]} f = \sum_{n=0}^{\infty} \rho^n \int \cdots \int_{t_1 < \cdots < t_n} \hat{f}(\{t_1, \ldots, t_n\}) \, \mathrm{d}B(t_1) \ldots \mathrm{d}B(t_n).$$

Convergence of operators follows from (2.4). The same holds for Example 3.7.

Example 5.2. A very different situation appears in Example 3.8. The second Brownian motion B_2 (or rather, its discrete approximation) is not linear but quadratic in random signs τ_m, $m \in \frac{1}{i}\mathbb{Z}$. It is two times less stable:

$$\mathbf{N}[i]f_{s,t}^{(2)}[i] = 2f_{s,t}^{(2)}[i]; \qquad \mathrm{Lim}_{i \to \infty}\, \rho^{\mathbf{N}[i]} = \rho^{2\mathbf{N}[\infty]},$$

if $\mathbf{N}[\infty]$ is defined in the same way as in Example 5.1. For B_3 it is $\rho^{3\mathbf{N}[\infty]}$, and so on. Still, $\sup_i \|f[i] - \rho^{\mathbf{N}[i]} f[i]\| \to 0$ for $\rho \to 1$. For B_λ, however, the change is dramatic. Namely,

$$\mathbf{N}[i]f_{s,t}^{(\lambda)}[i] = \mathrm{entier}(\lambda\sqrt{i})f_{s,t}^{(\lambda)}[i]; \qquad \mathrm{Lim}_{i \to \infty}\, \rho^{\mathbf{N}[i]} = 0^{\mathbf{N}[\infty]}$$

for all $\rho \in (-1,+1)$; here $0^{\mathbf{N}[\infty]} = \lim_{\rho \to 0} \rho^{\mathbf{N}[\infty]}$ is the orthogonal projection to the one-dimensional subspace of constants (just the expectation). The same holds for Example 3.9.

Notions of stability and sensitivity are introduced in [2, Sects. 1.1, 1.4] for a sequence of two-valued functions of $1, 2, 3, \ldots$ two-valued variables. For arbitrary (not just two-valued) functions, a number of equivalent definitions can be found in [12, Sect. 1]. They may be adapted to our framework as follows. We consider a function $f : \Omega[\mathrm{all}] \to \mathbb{R}$ such that $0 < \liminf_i \|f[i]\| \leq \limsup_i \|f[i]\| < \infty$. We say that f is stable, if $\sup_i \|f[i] - \rho^{\mathbf{N}[i]}f[i]\| \to 0$ when $\rho \to 1$. We say that f is sensitive, if $\|\rho^{\mathbf{N}[i]}f[i] - 0^{\mathbf{N}[i]}f[i]\| \to 0$ when $i \to \infty$, for some (therefore, every) $\rho \in (0,1)$. These definitions conform to [12] when $f[i]$ depends only on i signs $\tau_{1/i}, \ldots, \tau_{i/i}$. In terms of the two ρ-correlated arrays (τ_m), (τ_m'), stability means that $\mathbb{E}\left((f[i] \circ \alpha')(f[i] \circ \alpha)\right) \to \|f[i]\|^2$ for $\rho \to 1$, uniformly in i. Or, equivalently, $\mathbb{E}\left(\mathrm{Var}\left(f[i] \circ \alpha' \,\middle|\, \alpha^{-1}(\mathcal{F})\right)\right) \to 0$ when $\rho \to 1$, uniformly in i. Sensitivity means that $\mathbb{E}\left((f[i] \circ \alpha')(f[i] \circ \alpha)\right) \to \left(\mathbb{E}\, f[i]\right)^2$ when $n \to \infty$, for some (therefore, every) $\rho \in (0,1)$. Or, equivalently, $\mathbb{E}\left|\mathbb{E}\left(f[i] \circ \alpha' \,\middle|\, \alpha^{-1}(\mathcal{F})\right) - \mathbb{E}\, f[i]\right|^2 \to 0$ when $n \to \infty$, for some (therefore, every) $\rho \in (0,1)$.

In particular, those definitions can be applied to any $f \in L_2(\mathcal{A})$ such that $\|f[\infty]\| \neq 0$.

Example 5.1 shows that everything is stable in the Brownian coarse factorization. In contrast, everything is sensitive in the coarse factorization generated by B_λ in Example 5.2. In Sect. 5.3 we will find a reason to rename this 'stability' and 'sensitivity' as 'micro-stability' and 'micro-sensitivity'.

A sufficient condition for sensitivity is found by Benjamini, Kalai and Schramm in terms of the influence of a (two-valued) variable on a function, see [2, Sect. 1.2]. In our framework, the influence of the variable τ_m on a function $f[i] : \Omega[i] \to \mathbb{R}$ may be defined as the expectation of the square root of the conditional variance,

$$\mathbb{E}\,\sqrt{\mathrm{Var}\left(f[i] \,\middle|\, \mathcal{F}_{\frac{1}{i}\mathbb{Z}\setminus\{m\}}\right)};$$

here $\mathcal{F}_{\frac{1}{i}\mathbb{Z}\backslash\{m\}}$ is the sub-σ-field of $\mathcal{F}[i]$ generated by all random signs except for τ_m. The root of the conditional variance is simply one half of the difference between two values of the function $f[i]$, one value for $\tau_m = +1$, the other for $\tau_m = -1$. Thus, our formula gives two times less than [2, (1.3)], but the coefficient does not matter. Similarly, for any set $M \subset \frac{1}{i}\mathbb{Z}$, the influence of M (that is, of all variables τ_m, $m \in M$) on $f[i]$ may be defined as

$$\mathbb{E}\sqrt{\mathrm{Var}\left(f[i]\,\middle|\,\mathcal{F}_{\frac{1}{i}\mathbb{Z}\backslash M}\right)}.$$

By the way, for a *linear* function, the *squared* influence is additive (in M); indeed, if $f[i] = \sum_m c_m \tau_m$, then $\mathrm{Var}\left(f[i]\,\middle|\,\mathcal{F}_{\frac{1}{i}\mathbb{Z}\backslash M}\right) = \mathbb{E}\left(\sum_{m \in M} c_m \tau_m\right)^2 = \sum_{m \in M} c_m^2$. The sum of squared influences appears in the following remarkable result (adapted to our framework).

Theorem 5.3 (Benjamini, Kalai, Schramm). *Let a function $f : \Omega[\mathrm{all}] \to \{0, 1\}$ be such that each $f[i]$ depends on i variables $\tau_{1/i}, \ldots, \tau_{i/i}$ only. If*

$$\sum_{k=1}^{i}\left(\mathbb{E}\sqrt{\mathrm{Var}\left(f[i]\,\middle|\,\mathcal{F}_{\frac{1}{i}\mathbb{Z}\backslash\{k/i\}}\right)}\right)^2 \xrightarrow[i\to\infty]{} 0,$$

then f is sensitive.

See [2, Th. 1.3]. We will return to the point in Sect. 6.4.

5.2 Continuous Case

We start with the *Brownian* continuous factorization $\left((\Omega, \mathcal{F}, P), (\mathcal{F}_{s,t})_{s\le t}\right)$. Using the Wiener-Itô decomposition of $L_2(\Omega, \mathcal{F}, P)$,

$$f = \underbrace{\sum_{n=0}^{\infty}\int\cdots\int_{t_1<\cdots<t_n} \hat{f}(\{t_1, \ldots, t_n\})\,\mathrm{d}B(t_1)\ldots\mathrm{d}B(t_n)}_{\text{belongs to } n\text{-th Wiener chaos}}, \quad \hat{f} \in L_2(\mathcal{C}_{\mathrm{finite}}),$$

we can define a self-adjoint operator $\mathbf{N} : L_2 \to L_2$ such that for each n, $\mathbf{N}f = nf$ for all f of n-th Wiener chaos. Accordingly, $\rho^{\mathbf{N}}f = \rho^n f$ for these f. Informally, $\mathbf{N}(\mathrm{d}B(t_1)\ldots\mathrm{d}B(t_n)) = n\,\mathrm{d}B(t_1)\ldots\mathrm{d}B(t_n)$.

Every bounded Borel function φ on $\mathcal{C}_{\mathrm{finite}}$ acts on $L_2(\Omega, \mathcal{F}, P)$ by the operator R_φ,

$$R_\varphi f = \sum_{n=0}^{\infty}\int\cdots\int_{t_1<\cdots<t_n} \varphi(\{t_1, \ldots, t_n\})\hat{f}(\{t_1, \ldots, t_n\})\,\mathrm{d}B(t_1)\ldots\mathrm{d}B(t_n). \quad (5.1)$$

The operator $\rho^{\mathbf{N}}$ corresponds to the function $M \mapsto \rho^{|M|}$. (In some sense, the unbounded operator $\mathbf{N}$ corresponds to the unbounded function $M \mapsto |M|$.)

The decomposition $|M| = |M \cap (-\infty, t)| + |M \cap (t, \infty)|$ (it holds for μ_f-almost all M) leads to the operator decomposition $\mathbf{N} = \mathbf{N}_{-\infty,t} + \mathbf{N}_{t,\infty}$. Informally, $\mathbf{N}_{-\infty,t}(\mathrm{d}B(t_1)\dots\mathrm{d}B(t_n)) = k\mathrm{d}B(t_1)\dots\mathrm{d}B(t_n)$ and $\mathbf{N}_{t,\infty}(\mathrm{d}B(t_1)\dots\mathrm{d}B(t_n)) = (n-k)\mathrm{d}B(t_1)\dots\mathrm{d}B(t_n)$ whenever $t_1 < \dots < t_k < t < t_{k+1} < \dots < t_n$. Accordingly, $\rho^{\mathbf{N}} = \rho^{\mathbf{N}_{-\infty,t}} \otimes \rho^{\mathbf{N}_{t,\infty}}$.

A function $\varphi : \mathcal{C}_{\text{finite}} \to \{0,1\}$, the indicator of a Borel subset $\mathcal{M}$ of $\mathcal{C}_{\text{finite}}$, corresponds to the orthogonal projection operator onto the corresponding (recall Theorem 3.26) subspace $H_{\mathcal{M}}$. Say, for the (indicator of the) set $\{\emptyset\}$, the operator projects onto the one-dimensional space of constants (the expectation). For the set $\{M : M \subset (0,\infty)\}$ the operator is the conditional expectation, $\mathbb{E}\left(\cdot \mid \mathcal{F}_{0,\infty}\right)$.

The function

$$\varphi_{s,t}(M) = \begin{cases} 1 & \text{if } M \cap (s,t) \neq \emptyset, \\ 0 & \text{if } M \cap (s,t) = \emptyset \end{cases}$$

acts by the operator $\mathbf{1} - \mathbb{E}\left(\cdot \mid \mathcal{F}_{(-\infty,s)\cup(t,\infty)}\right)$.

For a finite set $L = \{s_1, \dots, s_n\} \subset \mathbb{R}$, $s_1 < \dots < s_n$, the function $\varphi_L(M) = \varphi_{s_1,s_2}(M) + \dots + \varphi_{s_{n-1},s_n}(M)$ counts intervals (s_j, s_{j+1}) that intersect M. Clearly, $\varphi_L(M) \leq |M|$, and

$$\varphi_{L_n}(M) \uparrow |M| \quad \text{for } \mu_f\text{-almost all } M$$

if $L_1 \subset L_2 \subset \dots$ are chosen so that their union is dense in $\mathbb{R}$. Accordingly,

$$\mathbf{N}_{L_n} \uparrow \mathbf{N},$$

$$\mathbf{N}_{\{s_1,\dots,s_n\}} = \sum_{j=1}^{n-1} \left(\mathbf{1} - \mathbb{E}\left(\cdot \mid \mathcal{F}_{(-\infty,s_j)\cup(s_{j+1},\infty)}\right)\right). \tag{5.2}$$

The operator $\mathbf{N}$ is thus expressed in terms of the factorization only, irrespective of the Wiener-Itô decomposition, which gives us a bridge to *arbitrary* continuous factorizations. Operators R_φ described in the next lemma generalize (5.1).

Lemma 5.4. *For every continuous factorization* $\left((\Omega, \mathcal{F}, P), (\mathcal{F}_{s,t})_{s \leq t}\right)$ *there exists one and only one map* $\varphi \mapsto R_\varphi$ *from the set of all bounded Borel functions* $\varphi : \mathcal{C} \to \mathbb{R}$ *to the set of (bounded linear) operators on* $L_2(\Omega, \mathcal{F}, P)$ *such that*

(a) the map is a homomorphism of algebras; that is, $R_{a\varphi} = aR_\varphi$, $R_{\varphi+\psi} = R_\varphi + R_\psi$, $R_{\varphi\psi} = R_\varphi R_\psi$;

(b) $\|R_\varphi\| \leq \sup_{M \in \mathcal{C}} |\varphi(M)|$;

(c) $R_{\mathbf{1}_{\mathcal{M}}} = \operatorname{Proj}_{H_{\mathcal{M}}}$ *for every Borel set* $\mathcal{M} \subset \mathcal{C}$; *here* $\mathbf{1}_{\mathcal{M}}$ *is the indicator of* $\mathcal{M}$, *and* $(H_{\mathcal{M}})$ *is the orthogonal decomposition provided by Theorem 3.26.*

The map also satisfies the condition

(d) let $\varphi, \varphi_1, \varphi_2, \dots : \mathcal{C} \to [0,1]$ *be Borel functions such that* $\varphi_k \to \varphi$ *pointwise (that is,* $\varphi_k(M) \xrightarrow[k\to\infty]{} \varphi(M)$ *for each* $M \in \mathcal{C}$); *then* $R_{\varphi_k} \to R_\varphi$ *strongly (that is,* $\|R_{\varphi_k}x - R_\varphi x\| \xrightarrow[k\to\infty]{} 0$ *for every* $x \in L_2(\Omega, \mathcal{F}, P)$).

Proof. Uniqueness and existence are easy: Condition (c) and linearity determine the map on the algebra of Borel functions $\varphi : \mathcal{C} \to \mathbb{R}$ having finite sets of values; it remains to extend the map by continuity.

For proving Condition (d) we note the equality

$$\langle R_\varphi x, x \rangle = \int \varphi \, d\mu_x \,,$$

where μ_x is the spectral measure of x; it holds for φ having finite sets of values, and therefore, for all φ. The bounded convergence theorem gives us not only $\langle R_{\varphi_k} x, x \rangle \to \langle R_\varphi x, x \rangle$, but also $\langle R_{(\varphi_k - \varphi)^2} x, x \rangle \to 0$. However, $\| R_{\varphi_k} x - R_\varphi x \|^2 = \langle R_{\varphi_k - \varphi} x, R_{\varphi_k - \varphi} x \rangle = \langle R_{(\varphi_k - \varphi)^2} x, x \rangle$. $\square$

Lemma 5.5. *For every continuous factorization* $((\Omega, \mathcal{F}, P), (\mathcal{F}_{s,t})_{s \leq t})$, *all finite sets* $L_1 \subset L_2 \subset \ldots$ *whose union is dense in* $\mathbb{R}$, *and every* $\lambda \in [0, \infty)$, *the limit*

$$U_\lambda = \lim_n \exp(-\lambda \mathbf{N}_{L_n}) \,,$$

where $\mathbf{N}_L$ *is defined by* (5.2), *exists in the strong operator topology, and does not depend on the choice of* $L_1, L_2, \ldots$ *Also,*

$$U_\lambda U_\mu = U_{\lambda + \mu} \quad \text{for all } \lambda, \mu \in [0, \infty) \,.$$

Proof. We have $\varphi_L = \sum \varphi_{s_k, s_{k+1}}$ and $R_{\varphi_{s,t}} = \mathbf{1} - \mathbb{E}\left(\cdot \,\middle|\, \mathcal{F}_{(-\infty,s) \cup (t,\infty)} \right)$; thus $R_{\varphi_L} = \mathbf{N}_L$. It follows that $R_{\exp(-\lambda \varphi_L)} = \exp(-\lambda \mathbf{N}_L)$. However, $\exp(-\lambda \varphi_{L_n}) \to \varphi_\lambda$, where $\varphi_\lambda(M) = \exp(-\lambda |M|)$ (and $e^{-\infty} = 0$, of course). By 5.4(d), $\exp(-\lambda \mathbf{N}_{L_n}) \to R_{\varphi_\lambda} = U_\lambda$. The semigroup relation $U_\lambda U_\mu = U_{\lambda + \mu}$ for operators follows from the corresponding relation $\varphi_\lambda \varphi_\mu = \varphi_{\lambda + \mu}$ for functions. $\square$

In the Brownian factorization we know that $U_\lambda = \exp(-\lambda \mathbf{N})$, $\mathbf{N} = \lim_n \mathbf{N}_{L_n}$. In general, however, the semigroup $(U_\lambda)_{\lambda \geq 0}$ is discontinuous at $\lambda = 0$ (and $\mathbf{N}$ is ill-defined).

Definition 5.6. *Let* $((\Omega, \mathcal{F}, P), (\mathcal{F}_{s,t})_{s \leq t})$ *be a continuous factorization, and* $f \in L_2(\Omega, \mathcal{F}, P)$.

(a) *f is called* stable, *if* $\| f - U_\lambda f \| \to 0$ *for* $\lambda \to 0$, *or equivalently, if* μ_f *is concentrated on* $\mathcal{C}_{\text{finite}} = \{ M \in \mathcal{C} : |M| < \infty \}$.

(b) *f is called* sensitive, *if* $U_\lambda f = 0$ *for all* $\lambda > 0$, *or equivalently, if* μ_f *is concentrated on* $\mathcal{C} \setminus \mathcal{C}_{\text{finite}} = \{ M \in \mathcal{C} : |M| = \infty \}$.

Of course, $U_0 f = f$ anyway. For proving equivalence, apply Lemma 5.4(d) to $U_\lambda = R_{\varphi_\lambda}$, $\varphi_\lambda(M) = e^{-\lambda |M|}$.

The space $L_2(\Omega, \mathcal{F}, P)$ decomposes into the direct sum of two subspaces, stable and sensitive, according to the decomposition of $\mathcal{C}$ into the union of two disjoint subsets, $\mathcal{C}_{\text{finite}}$ and $\mathcal{C} \setminus \mathcal{C}_{\text{finite}}$.

A continuous factorization is called *classical* (or *stable*), if the stable subspace is the whole $L_2(\Omega, \mathcal{F}, P)$.

A noise is called classical, if its continuous factorization is classical.

In order to understand probabilistic meaning of U_λ, consider first $\rho^{\mathbf{N}_L}$, $L = \{s_1, \ldots, s_n\}$, $s_1 < \cdots < s_n$. We have

$$\Omega = \Omega_{-\infty, s_1} \times \Omega_{s_1, s_2} \times \cdots \times \Omega_{s_{n-1}, s_n} \times \Omega_{s_n, \infty}$$

or rather, $(\Omega, \mathcal{F}, P) = (\Omega_{-\infty, s_1}, \mathcal{F}_{-\infty, s_1}, P_{-\infty, s_1}) \times \ldots$, but let me use the shorter notation. Each $\omega \in \Omega$ may be thought of as a sequence $(\omega_{-\infty, s_1}, \omega_{s_1, s_2}, \ldots \omega_{s_{n-1}, s_n}, \omega_{s_n, \infty})$ of local portions of data. Imagine another portion of data $\omega'_{s_1, s_2} \in \Omega_{s_1, s_2}$, either equal to ω_{s_1, s_2} (with probability ρ), or independent of it (with probability $1 - \rho$). The joint distribution of ω_{s_1, s_2} and ω'_{s_1, s_2} is a convex combination of two probability measures on $\tilde{\Omega}_{s_1, s_2} = \Omega_{s_1, s_2} \times \Omega_{s_1, s_2}$. One measure is concentrated on the diagonal and is the image of P_{s_1, s_2} under the map $\Omega_{s_1, s_2} \ni \omega_{s_1, s_2} \mapsto (\omega_{s_1, s_2}, \omega_{s_1, s_2}) \in \tilde{\Omega}_{s_1, s_2}$; this measure occurs with the coefficient ρ. The other measure is the product measure $P_{s_1, s_2} \otimes P_{s_1, s_2}$; it occurs with the coefficient $1 - \rho$.

Similarly we introduce $\tilde{\Omega}_{s_2, s_3}, \ldots, \tilde{\Omega}_{s_{n-1}, s_n}$ and construct $\tilde{\Omega} = \Omega_{-\infty, s_1} \times \tilde{\Omega}_{s_1, s_2} \times \cdots \times \tilde{\Omega}_{s_{n-1}, s_n} \times \Omega_{s_n, \infty}$ (the factors being equipped with corresponding measures). It is the same idea as in Sect. 5.1. Again, we have two measure preserving maps $\alpha, \alpha' : \tilde{\Omega} \to \Omega$. It appears that

$$\mathbb{E}\left(f \circ \alpha' \,\middle|\, \alpha^{-1}(\mathcal{F}) \right) = (\rho^{\mathbf{N}_L} f) \circ \alpha \quad \text{for } f \in L_2(\Omega, \mathcal{F}, P).$$

This is the probabilistic interpretation of $\rho^{\mathbf{N}_L}$; each portion of data is either correct (with probability ρ), or wrong (with probability $1 - \rho$).[30] However, the portions are not small yet. The limit $n \to \infty$ makes them infinitesimal, and turns $\rho^{\mathbf{N}_L}$ into U_λ, where ρ and λ are related by $\rho = e^{-\lambda}$.

The interpretation above motivates the terms 'stable' and 'sensitive'.

Constant functions on Ω are stable; sensitive functions are of zero mean. This is a terminological deviation from the discrete case; according to Sect. 5.1, constant functions are both stable and sensitive.

Two limiting cases of U_λ are projections. Namely, $U_\infty = \lim_{\lambda \to \infty} U_\lambda$ is the expectation, and $U_{0+} = \lim_{\lambda \to 0+} U_\lambda$ is the projection onto the stable subspace. Restricting the 'perturbation of local data' to a given interval (s, t) we get operators $U_\lambda^{(s,t)}$. These correspond to functions $\mathcal{C} \ni M \mapsto \exp(-\lambda |M \cap (s, t)|)$ and satisfy

---

[30] This time, $\rho \in [0, 1]$ rather than $[-1, 1]$. The relation to the approach of Sect. 5.1 is expressed by the equality

$$\frac{1 + \rho}{2} \begin{pmatrix} 1/2 & 0 \\ 0 & 1/2 \end{pmatrix} + \frac{1 - \rho}{2} \begin{pmatrix} 0 & 1/2 \\ 1/2 & 0 \end{pmatrix} = \begin{pmatrix} (1+\rho)/4 & (1-\rho)/4 \\ (1-\rho)/4 & (1+\rho)/4 \end{pmatrix}$$

$$= \rho \begin{pmatrix} 1/2 & 0 \\ 0 & 1/2 \end{pmatrix} + (1 - \rho) \begin{pmatrix} 1/4 & 1/4 \\ 1/4 & 1/4 \end{pmatrix}.$$

$$U_\lambda^{(s,t)} U_\mu^{(s,t)} = U_{\lambda+\mu}^{(s,t)}\,; \quad U_\lambda^{(r,s)} U_\lambda^{(s,t)} = U_\lambda^{(r,t)}\,;$$

$$U_\infty^{(s,t)} = \mathbb{E}\left(\,\cdot\,\big|\, \mathcal{F}_{-\infty,s} \otimes \mathcal{F}_{t,\infty}\right)\,; \tag{5.3}$$

$$U_{0+}^{(s,t)} = \mathbb{E}\left(\,\cdot\,\big|\, \mathcal{F}_{-\infty,s} \otimes \mathcal{F}_{s,t}^{\text{stable}} \otimes \mathcal{F}_{t,\infty}\right).$$

Note that (5.2) may be written as

$$\mathbf{N}_{\{s_1,\dots,s_n\}} = \left(1 - U_\infty^{(s_1,s_2)}\right) + \cdots + \left(1 - U_\infty^{(s_{n-1},s_n)}\right). \tag{5.4}$$

Lemma 5.7. *Let* $\left((\Omega,\mathcal{F},P),(\mathcal{F}_{s,t})_{s\leq t}\right)$ *be a continuous factorization,* $f \in L_2(\Omega,\mathcal{F},P)$, *and* $g = \eta \circ f$ *where* $\eta : \mathbb{R} \to \mathbb{R}$ *satisfies* $|\eta(x) - \eta(y)| \leq |x - y|$ *for all* $x, y \in \mathbb{R}$. *Then*

$$\mu_g(\mathcal{C} \setminus \mathcal{M}_E) \leq \mu_f(\mathcal{C} \setminus \mathcal{M}_E)$$

for all elementary sets $E \subset \mathbb{R}$; *here* $\mathcal{M}_E = \{M \in \mathcal{C} : M \subset E\}$.

Proof. We have (up to isomorphism) $\Omega = \Omega_E \times \Omega_{\mathbb{R}\setminus E}$ (the product of probability spaces is meant). We introduce $\tilde{\Omega} = \Omega \times \Omega = (\Omega_E \times \Omega_E) \times (\Omega_{\mathbb{R}\setminus E} \times \Omega_{\mathbb{R}\setminus E})$ and equip the second factor $\Omega_{\mathbb{R}\setminus E} \times \Omega_{\mathbb{R}\setminus E}$ with the product measure, while the first factor $\Omega_E \times \Omega_E$ is equipped with the measure concentrated on the diagonal, such that (equipping $\tilde{\Omega}$ with the product of these two measures), the measure preserving 'coordinate' maps $\alpha, \alpha' : \tilde{\Omega} \to \Omega$ satisfy

$$f \circ \alpha = f \circ \alpha' \quad \text{for all } \mathcal{F}_E\text{-measurable } f,$$
$$f \circ \alpha \text{ and } g \circ \alpha' \text{ are independent, for all } \mathcal{F}_{\mathbb{R}\setminus E}\text{-measurable } f, g.$$

Then

$$\mathbb{E}\left(f \circ \alpha' \,\big|\, \alpha^{-1}(\mathcal{F})\right) = \mathbb{E}\left(f \,\big|\, \mathcal{F}_E\right) \circ \alpha \quad \text{for all } f \in L_2(\Omega,\mathcal{F},P).$$

Therefore (recall Theorem 3.26),

$$\mathbb{E}\left((f \circ \alpha')(g \circ \alpha)\right) = \mathbb{E}\left(g\mathbb{E}\left(f \,\big|\, \mathcal{F}_E\right)\right);$$
$$\mathbb{E}\left((f \circ \alpha')(f \circ \alpha)\right) = \langle \operatorname{Proj}_{H_{\mathcal{M}_E}} f, f\rangle = \mu_f(\mathcal{M}_E);$$
$$\frac{1}{2}\mathbb{E}\left(f \circ \alpha' - f \circ \alpha\right)^2 = \mu_f(\mathcal{C}) - \mu_f(\mathcal{M}_E) = \mu_f(\mathcal{C} \setminus \mathcal{M}_E).$$

The same holds for g. It remains to note that $|g \circ \alpha' - g \circ \alpha| = |\eta \circ f \circ \alpha' - \eta \circ f \circ \alpha| \leq |f \circ \alpha' - f \circ \alpha|$ everywhere on $\tilde{\Omega}$. $\qquad\square$

We introduce a special set S of Borel functions $\varphi : \mathcal{C} \to [0,1]$ in three steps. First, we take all functions of the form $\mathbf{1}_{\mathcal{M}_E}$,

$$\mathbf{1}_{\mathcal{M}_E}(M) = \begin{cases} 1 & \text{if } M \subset E, \\ 0 & \text{otherwise,} \end{cases}$$

where $E \subset \mathbb{R}$ runs over all elementary sets. Second, we consider all (finite) convex combinations of these $\mathbf{1}_{M_E}$. Third, we consider the least set S containing these convex combinations and closed under pointwise convergence (that is, if $\varphi_k \in S$ and $\varphi_k(M) \to \varphi(M)$ for each $M \in \mathcal{C}$ then $\varphi \in S$).

The set S is convex (since the third step preserves convexity). It is also closed under multiplication: $\varphi\psi \in S$ for all $\varphi, \psi \in S$. Indeed, multiplicativity holds in the first step, and is preserved in the second and third steps.

Lemma 5.8. *Let $((\Omega, \mathcal{F}, P), (\mathcal{F}_{s,t})_{s \leq t})$ be a continuous factorization, $f \in L_2(\Omega, \mathcal{F}, P)$, and $g = \eta \circ f$ where $\eta : \mathbb{R} \to \mathbb{R}$ satisfies $|\eta(x) - \eta(y)| \leq |x - y|$ for all $x, y \in \mathbb{R}$. Then*

$$\int (1 - \varphi)\, d\mu_g \leq \int (1 - \varphi)\, d\mu_f$$

for all $\varphi \in S$.

Proof. In the first step, for $\varphi = \mathbf{1}_{M_E}$, the inequality is stated by Lemma 5.7. The second step evidently preserves the inequality. And the third step preserves it due to the bounded convergence theorem. $\quad\square$

Lemma 5.9. *Let a Borel set $\mathcal{M} \subset \mathcal{C}$ be such that its indicator function $\mathbf{1}_{\mathcal{M}}$ belongs to the set S. Then for every continuous factorization $((\Omega, \mathcal{F}, P), (\mathcal{F}_{s,t})_{s \leq t})$, the subspace $H_{\mathcal{M}} = \{f : \mu_f(\mathcal{C} \setminus \mathcal{M}) = 0\}$ of $L_2(\Omega, \mathcal{F}, P)$ is of the form*

$$H_{\mathcal{M}} = L_2(\Omega, \mathcal{F}_{\mathcal{M}}, P)$$

where $\mathcal{F}_{\mathcal{M}}$ is a sub-σ-field of $\mathcal{F}$.

Proof. The subspace satisfies

$$f \in H_{\mathcal{M}} \quad \text{implies} \quad |f| \in H_{\mathcal{M}}$$

(here $|f|(M) = |f(M)|$ for $M \in \mathcal{C}$). Indeed,

$$\int (1 - \mathbf{1}_{\mathcal{M}})\, d\mu_{|f|} \leq \int (1 - \mathbf{1}_{\mathcal{M}})\, d\mu_f$$

by Lemma 5.8; that is, $\mu_{|f|}(\mathcal{C} \setminus \mathcal{M}) \leq \mu_f(\mathcal{C} \setminus \mathcal{M})$. A subspace satisfying such a condition is necessarily of the form $L_2(\Omega, \mathcal{F}_{\mathcal{M}}, P)$. $\quad\square$

Recall the decomposition of $L_2(\Omega, \mathcal{F}, P)$ into the sum of two orthogonal subspaces, stable and sensitive, according to the decomposition of $\mathcal{C}$ into the union of two disjoint subsets, $\mathcal{C}_{\text{finite}}$ and $\mathcal{C} \setminus \mathcal{C}_{\text{finite}}$.

Theorem 5.10. *For every continuous factorization $((\Omega, \mathcal{F}, P), (\mathcal{F}_{s,t})_{s \leq t})$, there exists a sub-$\sigma$-field $\mathcal{F}_{\text{stable}}$ of $\mathcal{F}$ such that for all $f \in L_2(\Omega, \mathcal{F}, P)$*

$$f \text{ is stable if and only if } f \text{ is } \mathcal{F}_{\text{stable}}\text{-measurable};$$

$$f \text{ is sensitive if and only if } \mathbb{E}\left(f \mid \mathcal{F}_{\text{stable}}\right) = 0.$$

Proof. The second statement (about sensitive functions) follows from the first (about stable functions). By Lemma 5.9 it is enough to prove that the indicator of $\mathcal{C}_{\text{finite}}$ belongs to S.

For every $\lambda \in (0, \infty)$ the function $\varphi_\lambda : \mathcal{C} \to [0,1]$ defined by $\varphi_\lambda(M) = \exp(-\lambda|M|)$ belongs to S due to the limiting procedure $\varphi_\lambda = \lim \exp(-\lambda \varphi_{L_n})$ used in the proof of Lemma 5.5. For each n the function $\exp(-\lambda \varphi_{L_n}) = \prod \exp(-\lambda \varphi_{s_k, s_{k+1}})$ belongs to S, since each $\exp(-\lambda \varphi_{s,t})$ is a convex combination of two indicators, of $\mathcal{M}_{(-\infty, s) \cup (t, \infty)}$ and of the whole $\mathcal{M}$.

It remains to note that φ_λ converges for $\lambda \to 0$ to the indicator of $\mathcal{C}_{\text{finite}}$. $\qquad\square$

So, a continuous factorization (or a noise) is classical if and only if $\mathcal{F}_{\text{stable}} = \mathcal{F}$.

5.3 Back to Discrete: Two Kinds of Stability

The operator equality $\operatorname{Lim} \rho^{\mathbf{N}[i]} = \rho^{\mathbf{N}[\infty]}$ holds for some dyadic coarse factorizations (recall Example 5.1) but fails for some others (recall Example 5.2). Nothing like that happens for spectral measures; $\mu_f[i] \to \mu_f[\infty]$ always (see Theorem 3.14 and Sect. 3.4). However, the operator $\rho^{\mathbf{N}[i]}$ corresponds to the function $\mathcal{C}[i] \ni M \mapsto \rho^{|M|}$ treated as an element of $L_\infty(\mu_f[i])$, and the operator $\rho^{\mathbf{N}[\infty]}$ corresponds to the function $\mathcal{C}[\infty] \ni M \mapsto \rho^{|M|}$ treated as an element of $L_\infty(\mu_f[\infty])$. How is it possible? Where is the origin of the clash between discrete and continuous?

The origin is discontinuity of functions $M \mapsto \rho^{|M|}$ and $M \mapsto |M|$ w.r.t. the Hausdorff topology on $\mathcal{C}$.

Example 5.11. Return to the equality $\mathbf{N}[i] f_{s,t}^{(2)}[i] = 2 f_{s,t}^{(2)}[i]$ for $f_{s,t}^{(2)}[i] = i^{-1/2} \sum \tau_m \tau_{m+(1/i)}$ (see Examples 5.2 and 3.8). The spectral measure of $f_{s,t}^{(2)}[i]$ is concentrated on two-point sets $M \subset \frac{1}{i}\mathbb{Z}$, namely, on pairs of two adjacent points $\{m, m + (1/i)\}$. However, $f_{s,t}^{(2)}[\infty]$ is just a Brownian increment; its spectral measure is concentrated on single-point sets. Now we see what happens; two close points merge in the limit! Multiplicity of spectral points eludes the continuous model.

The effect becomes dramatic for $f_{s,t}^{(\lambda)}[i]$; everything is stable in the continuous model ($i = \infty$), while everything is sensitive (for $i \to \infty$) in the discrete model. A finite spectral set on the continuum hides the infinite multiplicity of each point.

Conformity between discrete and continuous can be restored by modifying the idea of stability introduced in Sect. 5.1. Instead of inverting each τ_m (with probability $(1 - \rho)/2$ independently of others), we may invert blocks $\tau_{s[i]}, \tau_{s[i]+(1/i)}, \dots, \tau_{t[i]}$ where coarse instants s, t satisfy $t[\infty] - s[\infty] = \varepsilon$. Each block is inverted with probability $(1 - \rho)/2$, independently of other blocks. Ultimately we let $\varepsilon \to 0$, but the order of limits is crucial: $\lim_{\varepsilon \to 0} \lim_{i \to \infty}(\dots)$.

This way, we can define (in discrete time setup) *block stability* and *block sensitivity*, equivalent to stability and sensitivity (resp.) of the refinement. In contrast, the approach of Sect. 5.1 leads to what may be called *micro-stability* and *micro-sensitivity* (for discrete time only).

The function $\mathcal{C} \ni M \mapsto \rho^{|M|}$ is not continuous, but it is upper semicontinuous. Therefore, every micro-stable function is block stable, and every block sensitive function is micro-sensitive.

Example 5.12. The function $g_{s,t}$ of Example 3.9 is micro-sensitive but block stable. The same holds for all coarse random variables in that dyadic coarse factorization. It holds also for the second construction of Example 3.8 (I mean $f_{s,t}^{(\lambda)}$).

6 Generalizing Wiener Chaos

6.1 First Chaos, Decomposable Processes, Stability

We consider an arbitrary continuous factorization. As was shown in Theorem 3.26 and Lemma 5.4, Borel functions $\varphi : \mathcal{C} \to \mathbb{R}$ act on $L_2(\Omega, \mathcal{F}, P)$ by linear operators R_φ, and (indicators of) Borel subsets $\mathcal{M} \subset \mathcal{C}$ act by orthogonal projections to subspaces $H_\mathcal{M}$.

In particular, for the Brownian factorization, only $\mathcal{C}_{\text{finite}}$ is relevant. The set $\{M \in \mathcal{C}_{\text{finite}} : |M| = n\}$ corresponds to the subspace called n-th Wiener chaos.

In general, we may define n-th chaos as the subspace of $L_2(\Omega, \mathcal{F}, P)$ that corresponds to $\{M \in \mathcal{C} : |M| = n\}$. These subspaces are orthogonal, and span the stable subspace — not the whole $L_2(\Omega, \mathcal{F}, P)$, unless the noise is classical.

For each $t \in \mathbb{R}$ the set $\mathcal{M}_t = \{M : M \ni t\}$ is negligible in the sense that $H_{\mathcal{M}_t} = \{0\}$ (recall Lemma 3.15 and (3.11)). Neglecting $\mathcal{M}_t$ we may treat $\mathcal{C}$ as the product,[31]

$$\mathcal{C} = \mathcal{C}_{-\infty,t} \times \mathcal{C}_{t,\infty} , \tag{6.1}$$

where $\mathcal{C}_{a,b}$ is the space of all compact subsets of (a,b); namely, we treat a set $M \in \mathcal{C}$ as the pair of sets $M \cap (-\infty, t)$ and $M \cap (t, \infty)$, assuming $t \notin M$.

On the other hand, the Hilbert space $H = H_\mathcal{C} = L_2(\Omega, \mathcal{F}, P)$ may be treated as the tensor product,

$$H = H_{-\infty,t} \otimes H_{t,\infty} ,$$

of two Hilbert spaces $H_{-\infty,t} = H_{\mathcal{C}_{-\infty,t}} = L_2(\Omega, \mathcal{F}_{-\infty,t}, P)$ and $H_{t,\infty} = H_{\mathcal{C}_{t,\infty}} = L_2(\Omega, \mathcal{F}_{t,\infty}, P)$. Namely, $f \otimes g$ is just the usual product fg of random variables $f \in L_2(\Omega, \mathcal{F}_{-\infty,t}, P)$ and $g \in L_2(\Omega, \mathcal{F}_{t,\infty}, P)$; note that f and g are necessarily independent, therefore $\mathbb{E}|fg|^2 = (\mathbb{E}|f|^2)(\mathbb{E}|g|^2)$.

Subspaces $H_\mathcal{M} \subset H_{-\infty,t}$ for Borel subsets $\mathcal{M} \subset \mathcal{C}_{-\infty,t}$ are a σ-additive orthogonal decomposition of $H_{-\infty,t}$. The same holds for (t, ∞).

Lemma 6.1. $H_{\mathcal{M}_1 \times \mathcal{M}_2} = H_{\mathcal{M}_1} \otimes H_{\mathcal{M}_2}$ *for all Borel sets* $\mathcal{M}_1 \subset \mathcal{C}_{-\infty,t}$ *and* $\mathcal{M}_2 \subset \mathcal{C}_{t,\infty}$.

Proof. The equality holds for the special case $\mathcal{M}_1 = \{M : M \subset E_1\}$, $\mathcal{M}_2 = \{M : M \subset E_2\}$ where $E_1 \subset (-\infty, t)$ and $E_2 \subset (t, \infty)$ are elementary sets; indeed, $L_2(\Omega, \mathcal{F}_{E_1}, P) \otimes L_2(\Omega, \mathcal{F}_{E_2}, P) = L_2(\Omega, \mathcal{F}_{E_1 \cup E_2}, P)$ since $\mathcal{F}_{E_1 \cup E_2} = \mathcal{F}_{E_1} \otimes \mathcal{F}_{E_2}$. The general case follows by the monotone class theorem. $\square$

Theorem 6.2. *The sub-σ-field generated by the first chaos is equal to* $\mathcal{F}_{\text{stable}}$.

---

[31] Sorry, the formula '$\mathcal{C} = \mathcal{C}_{-\infty,t} \times \mathcal{C}_{t,\infty}$' may be confusing since, on the other hand, $\mathcal{C}_{-\infty,t} \subset \mathcal{C}$ and $\mathcal{C}_{t,\infty} \subset \mathcal{C}$. The same can be said about the next formula, $H = H_{-\infty,t} \otimes H_{t,\infty}$.

Proof. The σ-field is evidently included in $\mathcal{F}_{\text{stable}}$. Given a finite set $L = \{s_1, \ldots, s_n\} \subset \mathbb{R}$, $s_1 < \cdots < s_n$, we consider the set $\mathcal{M}_L$ of all $M \in \mathcal{C}$ such that $M \subset (s_1, s_n)$ and each $[s_k, s_{k+1}]$ contains at most one point of M. The set $\mathcal{M}_L$ being the product (over k), Lemma 6.1 shows that $H_{\mathcal{M}_L}$ is the tensor product (over k) of subspaces of $L_2(\Omega, \mathcal{F}_{s_k, s_{k+1}}, P)$; each factor is the first chaos on (s_k, s_{k+1}) plus constants. Therefore each function of $H_{\mathcal{M}_L}$ is measurable w.r.t. the σ-field generated by the first chaos. We choose $L_1 \subset L_2 \subset \ldots$ whose union is dense in $\mathbb{R}$; then $\mathcal{M}_{L_n} \uparrow \mathcal{C}_{\text{finite}}$, and corresponding subspaces span the stable subspace. $\qquad\square$

A random variable $X \in L_2(\Omega, \mathcal{F}, P)$ belongs to the first chaos if and only if

$$X = \mathbb{E}\left(X \,\middle|\, \mathcal{F}_{-\infty, t}\right) + \mathbb{E}\left(X \,\middle|\, \mathcal{F}_{t, \infty}\right) \quad \text{for all } t \in \mathbb{R}.$$

For such X, letting $X_{s,t} = \mathbb{E}\left(X \,\middle|\, \mathcal{F}_{s,t}\right)$ we get a *decomposable process*, that is, a family $(X_{s,t})_{s \leq t}$ of random variables such that $X_{s,t}$ is $\mathcal{F}_{s,t}$-measurable and $X_{r,s} + X_{s,t} = X_{r,t}$ whenever $r \leq s \leq t$. This way we get decomposable processes satisfying $\mathbb{E}|X_{s,t}|^2 < \infty$ and $\mathbb{E} X_{s,t} = 0$. Waiving these additional conditions we get a larger set of processes, but the sub-σ-field generated by these processes is still $\mathcal{F}_{\text{stable}}$. We may also consider complex-valued *multiplicative decomposable processes;* it means that $X_{s,t} : \Omega \to \mathbb{C}$ is $\mathcal{F}_{s,t}$-measurable and $X_{r,s} X_{s,t} = X_{r,t}$. The generated sub-$\sigma$-field is $\mathcal{F}_{\text{stable}}$, again. The same holds under the restriction $|X_{s,t}| = 1$ a.s. See [20, Th. 1.7].

Dealing with a noise (rather than factorization) we may restrict ourselves to stationary Brownian and Poisson decomposable processes. 'Stationary' means $X_{r,s} \circ \alpha_t = X_{r-t, s-t}$. 'Brownian' means $X_{s,t} \sim \mathrm{N}(0, t-s)$. 'Poisson' means $X_{s,t} \sim \mathrm{Poisson}(\lambda(t-s))$ for some $\lambda \in (0, \infty)$. The generated sub-σ-field is still $\mathcal{F}_{\text{stable}}$. See [15, Lemma 2.9]. (It was written for the Brownian component, but works also for the Poisson component.)

For a finite set $L = \{s_1, \ldots, s_n\} \subset \mathbb{R}$, $s_1 < \cdots < s_n$, we introduce an operator Q_L on the space $L_2^0 = \{X \in L_2(\Omega, \mathcal{F}, P) : \mathbb{E} X = 0\}$ by

$$Q_L = \mathbb{E}\left(\cdot \,\middle|\, \mathcal{F}_{-\infty, s_1}\right) + \mathbb{E}\left(\cdot \,\middle|\, \mathcal{F}_{s_1, s_2}\right) + \cdots + \mathbb{E}\left(\cdot \,\middle|\, \mathcal{F}_{s_{n-1}, s_n}\right) + \mathbb{E}\left(\cdot \,\middle|\, \mathcal{F}_{s_n, \infty}\right).$$

Theorem 6.3. *If finite sets $L_1 \subset L_2 \subset \ldots$ are such that their union is dense in $\mathbb{R}$, then operators Q_{L_n} converge in the strong operator topology to the orthogonal projection from L_2^0 onto the first chaos.*

Proof. Q_L is the projection onto $H_{\mathcal{M}_L}$, where $\mathcal{M}_L$ is the set of all nonempty $M \in \mathcal{C}$ contained in one of the $n + 1$ intervals. The intersection of subspaces corresponds to the intersection of subsets. $\qquad\square$

Stochastic analysis gives us another useful tool for calculating the first chaos, pioneered by Jon Warren [23, Th. 12]. Let $(B_{s,t})_{s \leq t}$ be a decomposable Brownian motion, that is, a decomposable process such that $B_{s,t} \sim \mathrm{N}(0, t-s)$. One says that B has the *representation property*, if every $X \in L_2(\Omega, \mathcal{F}, P)$ such that $\mathbb{E} X = 0$ is equal to a stochastic integral,

$$X = \int_{-\infty}^{+\infty} H(t)\, \mathrm{d}B_{0,t}\,,$$

where H is a predictable *process* w.r.t. the filtration $(\mathcal{F}_{-\infty,t})_{t\in\mathbb{R}}$.

Lemma 6.4. *If B has the representation property then the first chaos is equal to the set of all* linear *stochastic integrals*

$$\int_{-\infty}^{+\infty} \varphi(t)\, \mathrm{d}B_{0,t}\,, \qquad \varphi \in L_2(\mathbb{R})\,.$$

Proof. Linear stochastic integrals evidently belong to the first chaos. Let X belong to the first chaos. Consider martingales $B(t) = B_{0,t}$, $X(t) = \mathbb{E}\left(X\,|\,\mathcal{F}_{-\infty,t}\right) = \int_{-\infty}^{t} H(s)\,\mathrm{d}B(s)$ and their bracket process $\langle X, B \rangle_t = \int_{-\infty}^{t} H(s)\,\mathrm{d}s$. The two-dimensional process $(B(\cdot), X(\cdot))$ has independent increments; therefore the bracket process has independent increments as well. On the other hand, the bracket process is a continuous process of finite variation. Therefore it is degenerate (non-random), and $H(\cdot)$ is also non-random. □

It follows that $\mathcal{F}_{\mathrm{stable}}$ is generated by B.

Example 6.5. For the noise of stickiness (see Sect. 4), the process $(a(s,t))_{s\le t}$ is a decomposable Brownian motion having the representation property. Therefore it generates $\mathcal{F}_{\mathrm{stable}}$. On the other hand we know (recall Example 4.9) that $a(\cdot,\cdot)$ does not generate the whole σ-field. So, the sticky noise is not classical (Warren [23]).

The approach of Theorem 6.3 is also applicable. Let $\varphi : G_3 \to [-1,+1]$ be a Borel function, and $0 < t - \varepsilon < t < 1$. We consider $\varphi(\xi_{0,1}) = \varphi(\xi_{0,t-\varepsilon}\xi_{t-\varepsilon,t}\xi_{t,1})$ (you know, $\xi_{t-\varepsilon,t} = f_{a(t-\varepsilon,t),b(t-\varepsilon,t),c(t-\varepsilon,t)}$), and compare it with $\varphi(\xi_{0,t-\varepsilon}\tilde{\xi}_{t-\varepsilon,t}\xi_{t,1})$, where $\tilde{\xi}_{t-\varepsilon,t} = f_{a(t-\varepsilon,t),b(t-\varepsilon,t),0}$.

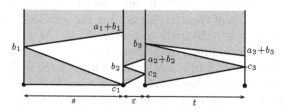

It appears that

$$\|\varphi(\xi_{0,t-\varepsilon}\xi_{t-\varepsilon,t}\xi_{t,1}) - \varphi(\xi_{0,t-\varepsilon}\tilde{\xi}_{t-\varepsilon,t}\xi_{t,1})\|_{L_2} = O(\varepsilon^{3/4}) = o(\sqrt{\varepsilon})\,,$$

provided that t is bounded away from 1 (otherwise we get $O(\varepsilon^{3/4}(1-t)^{-1/2}$ with an absolute constant). Taking into account that $\tilde{\xi}_{t-\varepsilon,t}$ is measurable w.r.t. the σ-field generated by $a(\cdot,\cdot)$ we conclude that the projection of $\varphi(\xi_{0,1})$ onto the first chaos is measurable w.r.t. the σ-field generated by $a(\cdot,\cdot)$. See Sect. 7.2 for the rest.

6.2 Higher Levels of Chaos

We still consider an arbitrary continuous factorization. Any Borel subset $\mathcal{M} \subset \mathcal{C}$ determines a subspace $H_\mathcal{M} \subset L_2(\Omega, \mathcal{F}, P)$. However, the subset $\mathcal{C}_{\text{finite}} \subset \mathcal{C}$ is special; the corresponding subspace, being equal to $L_2(\mathcal{F}_{\text{stable}})$ by Theorem 5.10, is of the form $L_2(\mathcal{F}_1)$ for a sub-σ-field $\mathcal{F}_1 \subset \mathcal{F}$.

Another interesting subset is $\mathcal{C}_{\text{countable}}$, the set of all at most countable compact subsets of $\mathbb{R}$. It is not a Borel subset of $\mathcal{C}$ [7, Th. 27.5] but still, it is universally measurable [7, Th. 21.10] (that is, measurable w.r.t. every Borel measure), since its complement is analytic [7, Th. 27.5]. The Cantor-Bendixson derivative M' of $M \in \mathcal{C}$ is, by definition, the set of all limit points of M. Clearly, $M' \in \mathcal{C}$, $M' \subset M$, and $M' = \emptyset$ if and only if M is finite. The iterated Cantor-Bendixson derivative $M^{(\alpha)}$ is defined for every ordinal α by transfinite recursion: $M^{(0)} = M$; $M^{(\alpha+1)} = (M^{(\alpha)})'$; and $M^{(\alpha)} = \cap_{\beta < \alpha} M^{(\beta)}$ if α is a limit ordinal; see [7, Sect. 6.C]. If $M \notin \mathcal{C}_{\text{countable}}$ then $M^{(\alpha)} \neq \emptyset$ for all α. If $M \in \mathcal{C}_{\text{countable}}$ then $M^{(\alpha)} = \emptyset$ for some finite or countable ordinal α; the least α such that $M^{(\alpha)} = \emptyset$ is called the Cantor-Bendixson rank of $M \in \mathcal{C}_{\text{countable}}$. It is always of the form $\beta + 1$, and $M^{(\beta)}$ is a finite set.

Recall the proof of Theorem 5.10: the indicator of $\mathcal{C}_{\text{finite}}$ belongs to the set S introduced in Sect. 5.2. Here is a more general fact.

Lemma 6.6. *Let α be an at most countable ordinal, and $\mathcal{M}_\alpha$ the set of all $M \in \mathcal{C}$ such that $M^{(\alpha)} = \emptyset$. Then the indicator function of $\mathcal{M}_\alpha$ belongs to the set S.*

Proof. Transfinite induction in α. For $\alpha = 0$ the claim is trivial. Let α be a limit ordinal. We take $\alpha_k \uparrow \alpha$, $\alpha_k < \alpha$, and note that $\mathcal{M}_\alpha = \mathcal{M}_{\alpha_1} \cup \mathcal{M}_{\alpha_2} \cup \ldots$ (indeed, $M^{(\alpha_k)} \downarrow M^{(\alpha)}$, and $M^{(\alpha_k)}$ are compact). Thus, indicators of $\mathcal{M}_{\alpha_k}$ converge to the indicator of $\mathcal{M}_\alpha$.

The transition from α to $\alpha + 1$ needs the following property of S: for every $\varphi \in S$ and a closed elementary set E, the function $M \mapsto \varphi(M \cap E)$ belongs to S. Proof: In the first step of constructing S, φ is the indicator of some $\{M : M \subset E_1\}$; thus $M \mapsto \varphi(M \cap E)$ is the indicator of $\{M : M \subset E_1 \cup (\mathbb{R} \setminus E)\}$. The second and third steps preserve the property.

Assume that the indicator function of $\mathcal{M}_\alpha$ belongs to S; we have to prove the same for $\alpha + 1$. The indicator of $\mathcal{M}_{\alpha+1}$ is $M \mapsto \varphi(M^{(\alpha)})$, where φ is the indicator of $\mathcal{C}_{\text{finite}}$. Taking into account that $\varphi \in S$ (see the proof of Theorem 5.10), we will prove a more general fact: the function $M \mapsto \varphi(M^{(\alpha)})$ belongs to S for every $\varphi \in S$ (not just the indicator of $\mathcal{C}_{\text{finite}}$). The property is evidently preserved by the second and third steps of constructing S; it remains to prove it in the first step. Here φ is the indicator of $\{M : M \subset E\}$ for an elementary E. We have to express the set $\{M : M^{(\alpha)} \subset E\}$ as a limit of sets of the form $\{M : (M \cap E_1)^{(\alpha)} = \emptyset\}$ where E_1 is a closed elementary set. The indicator of $\{M : (M \cap E_1)^{(\alpha)} = \emptyset\}$ belongs to S, since it is $\mathbf{1}_{\mathcal{M}_\alpha}(M \cap E_1)$. We note that, for $\varepsilon \to 0$,

$$\{M : (M \cap (-\infty, \varepsilon])^{(\alpha)} = \emptyset\} \uparrow \{M : M^{(\alpha)} \subset (0, \infty)\},$$
$$\{M : (M \cap (-\infty, -\varepsilon])^{(\alpha)} = \emptyset\} \downarrow \{M : M^{(\alpha)} \subset [0, \infty)\},$$

which does the job for two special cases, $E = (0, \infty)$ and $E = [0, \infty)$, and shows how to deal with a boundary point, belonging to E or not. The general case is left to the reader. □

Theorem 6.7. *Let $\big((\Omega, \mathcal{F}, P), (\mathcal{F}_{s,t})_{s \leq t}\big)$ be a continuous factorization.*
   (a) *There exists a sub-σ-field $\mathcal{E}$ of $\mathcal{F}$ such that for all $f \in L_2(\Omega, \mathcal{F}, P)$, f is $\mathcal{E}$-measurable if and only if μ_f is concentrated on $C_{\text{countable}}$.*
   (b) *For every at most countable ordinal α there exists a sub-σ-field $\mathcal{E}_\alpha$ of $\mathcal{F}$ such that for all $f \in L_2(\Omega, \mathcal{F}, P)$, f is $\mathcal{E}_\alpha$-measurable if and only if μ_f is concentrated on the set of $M \in \mathcal{C}$ such that $M^{(\alpha)} = \emptyset$ (that is, of Cantor-Bendixson rank less than or equal to α).*

Proof. Item (a) follows from (b), since $\mathcal{E}_\alpha = \mathcal{E}_{\alpha+1}$ for countable α large enough (see [7, Th. 6.9]), and $\mu_f(C_{\text{countable}}) = \sup_\alpha \mu_f\{M : M^{(\alpha)} = \emptyset\}$ (see [7], the proof of Th. 21.10, and Th. 35.23).
   Item (b) follows from Lemmas 6.6, 5.9. □

Let us concentrate on Item (b) for $\alpha = 0, 1, 2$. The case $\alpha = 0$ is trivial: only the empty set M, and only constant functions f. The case $\alpha = 1$ was discussed before: finite sets M and stable functions f. The case $\alpha = 2$ means that M' is finite.
   We define the n-th *superchaos* as the subspace $H_\mathcal{M} \subset L_2(\Omega, \mathcal{F}, P)$ corresponding to $\{M \in \mathcal{C} : |M'| = n\}$. These subspaces are orthogonal. The 0-th superchaos is the stable subspace, while for $n = 1, 2, \ldots$ the n-th superchaos consists of (some) sensitive functions. By Theorem 6.7(b), the subspace spanned by n-th superchaos spaces for all $n = 0, 1, 2, \ldots$ is of the form $L_2(\Omega, \mathcal{E}_2, P)$ where $\mathcal{E}_2$ is a sub-σ-field of $\mathcal{F}$. Similarly to Theorem 6.2, the sub-σ-field generated by the first superchaos and $\mathcal{F}_{\text{stable}}$ is equal to $\mathcal{E}_2$.
   Similarly to (5.2) and (5.4) we may 'count' points of M' by the operator

$$\mathbf{N}'_{\{s_1, \ldots, s_n\}} = \sum_{j=1}^{n-1} \big(1 - \mathbb{E}\big(\,\cdot\,\big|\, \mathcal{F}_{-\infty, s_j} \otimes \mathcal{F}^{\text{stable}}_{s_j, s_{j+1}} \otimes \mathcal{F}_{s_{j+1}, \infty}\big)\big)$$
$$= \big(1 - U_{0+}^{(s_1, s_2)}\big) + \cdots + \big(1 - U_{0+}^{(s_{n-1}, s_n)}\big),$$

or rather its limit $\mathbf{N}' = \lim_n \mathbf{N}'_{L_n}$. Further, similarly to Lemma 5.5, we may define

$$V_\lambda = \lim_n \exp(-\lambda \mathbf{N}'_{L_n}).$$

This way, an ordinal hierarchy of operators may be constructed. It corresponds to the Cantor-Bendixson hierarchy of countable compact sets.

Introducing

$$Q'_{\{s_1,\ldots,s_n\}}X = \mathbb{E}\left(X\mid \mathcal{F}_{-\infty,s_1}\otimes\mathcal{F}_{s_1,\infty}^{\text{stable}}\right)+\mathbb{E}\left(X\mid \mathcal{F}_{-\infty,s_1}^{\text{stable}}\otimes\mathcal{F}_{s_1,s_2}\otimes\mathcal{F}_{s_2,\infty}^{\text{stable}}\right)$$
$$+\cdots+\mathbb{E}\left(X\mid \mathcal{F}_{-\infty,s_{n-1}}^{\text{stable}}\otimes\mathcal{F}_{s_{n-1},s_n}\otimes\mathcal{F}_{s_n,\infty}^{\text{stable}}\right)+\mathbb{E}\left(X\mid \mathcal{F}_{-\infty,s_n}^{\text{stable}}\otimes\mathcal{F}_{s_n,\infty}\right)$$

for $X \in L_2(\Omega,\mathcal{F},P)$ such that $\mathbb{E}\left(X\mid \mathcal{F}_{\text{stable}}\right) = 0$, we get such a counterpart of Theorem 6.3.

Theorem 6.8. *If finite sets $L_1 \subset L_2 \subset \ldots$ are such that their union is dense in $\mathbb{R}$, then operators Q'_{L_n} converge in the strong operator topology to the orthogonal projection from the sensitive subspace onto the first superchaos.*

Proof. Q'_L is the projection onto $H_{\mathcal{M}_L}$, where $\mathcal{M}_L$ is the set of all nonempty $M \in \mathcal{C}$ such that M' is contained in one of the $n+1$ intervals. The intersection of subspaces corresponds to the intersection of subsets. □

Example 6.9. For the sticky noise, consider such a random variable X: the number of random chords $[s,t] \times \{x\}$ such that $s > 0$ and $t > 1$. In other words (see Sect. 4.9),
$$X = |\{x : \sigma_1(x) \in \Pi \cap (0,\infty)\}|.$$
The conditional distribution of X given the Brownian path $B(\cdot) = a(0,\cdot)$ is Poisson(λ) with $\lambda = a(0,1) + b(0,1) = B(1) - \min_{[0,1]} B(\cdot)$, which is easy to guess from the discrete counterpart (see (4.13)). That is a generalization of a claim from Example 4.9. In fact, the conditional distribution of the set $\{x : \sigma_1(x) \in \Pi \cap (0,\infty)\}$, given the Brownian path, is the Poisson point process of intensity 1 on $[-b(0,1), a(0,1)]$, which is a result of Warren [23]. Taking into account that the σ-field generated by $B(\cdot)$ is $\mathcal{F}_{\text{stable}}$ (recall Lemma 6.5), we get $\mathbb{E}\left(X\mid \mathcal{F}_{\text{stable}}\right) = a(0,1) + b(0,1)$. The random variable
$$Y = X - \mathbb{E}\left(X\mid \mathcal{F}_{\text{stable}}\right) = X - a(0,1) - b(0,1)$$
is sensitive, that is, $\mathbb{E}\left(Y\mid \mathcal{F}_{\text{stable}}\right) = 0$. I claim that Y belongs to the first superchaos.

The proof is based on Theorem 6.8. Given $0 < s_1 < \cdots < s_n < 1$, we have to check that Y can be decomposed into a sum $Y_0 + \cdots + Y_n$ such that each Y_j is measurable w.r.t. $\mathcal{F}_{0,s_j}^{\text{stable}} \otimes \mathcal{F}_{s_j,s_{j+1}} \otimes \mathcal{F}_{s_{j+1},1}^{\text{stable}}$. Here is the needed decomposition:
$$X_j = |\{x : \sigma_1(x) \in \Pi \cap (s_j, s_{j+1})\}|,$$
$$Y_j = X_j - \mathbb{E}\left(X_j\mid \mathcal{F}_{\text{stable}}\right).$$

We apply a small perturbation on $(0, s_j)$ and $(s_{j+1}, 1)$ but not on (s_j, s_{j+1}). The set $\Pi \cap (s_j, s_{j+1})$ remains unperturbed. The function σ_1 is perturbed, but only a little; being a function of $B(\cdot)$, it is stable.

So, Y belongs to the first superchaos, and X belongs to the first superchaos plus $L_2(\mathcal{F}_{\text{stable}})$. It means that μ_X is concentrated on sets M such that $|M'| \leq 1$.

The same holds for random variables $X_u = |\{x : x \le u, \sigma_1(x) \in \Pi \cap (0, \infty)\}|$, for any u. They all are measurable w.r.t. the σ-field generated by the first superchaos and $\mathcal{F}_{\text{stable}}$. The random variable $c(0, 1)$ is a (nonlinear!) function of these X_u (recall Sect. 4.9). We see that the first superchaos and $\mathcal{F}_{\text{stable}}$ generate the whole σ-field $\mathcal{F}$. Every spectral set (of every random variable) has only a finite number of limit points.

Example 6.10. Another nonclassical noise, discovered and investigated by Warren [22], see also Watanabe [25], may be called the noise of splitting. It is the scaling limit of the model of Example 1.9; see also Sect. 8.3. Spectral measures of the most interesting random variables are described explicitly! A spectral set contains a single limit point, and two sequences converging to the point from the left and from the right.

Again, every spectral set (of every random variable) has only a finite number of limit points.

Question 6.11. We have no example of a noise whose spectral sets M are at most countable, and M' is not always finite. Can it happen at all? Can it happen for the refinement of a dyadic coarse factorization satisfying (3.13)?

Beyond $\mathcal{C}_{\text{countable}}$ it is natural to use the Hausdorff dimension, $\dim M$, of compact sets $M \in \mathcal{C}$. The set S used in Theorems 5.10 and 6.7 helps again. First, a general lemma.

Lemma 6.12. *For every probability measure μ on $\mathcal{C}$ the function $\varphi : \mathcal{C} \to [0, 1]$ defined by $\varphi(M) = \mu\{M_1 \in \mathcal{C} : M \cap M_1 = \emptyset\}$, belongs to the set S.*

Proof. We may restrict ourselves to compact subsets of a bounded interval; let it be just $[0, 1]$. For any such set M let $M^{(n)}$ denote the union of intervals $[\frac{k}{n}, \frac{k+1}{n}]$ $(k = 0, \ldots, n - 1)$ that intersect M. The sequence $(M^{(n)})_{n=1}^{\infty}$ decreases and converges to M (in the Hausdorff metric). For every n, the function $\varphi_n(M) = \mu\{M_1 : M \cap M_1^{(n)} = \emptyset\}$ belongs to S, since it is the convex combination of indicators of $\{M : M \subset E\}$ with coefficients $\mu\{M_1 : M_1^{(n)} = [0, 1] \setminus E\}$, where E runs over 2^n elementary sets. It remains to note that $\varphi_n(M) \uparrow \varphi(M)$, since $M \cap M_1 = \emptyset$ if and only if $M \cap M_1^{(n)} = \emptyset$ for some n. $\square$

Lemma 6.13. *For every $\alpha \in (0, 1)$ there exists a function $\varphi \in S$ such that $\varphi(M) = 1$ for all M satisfying $\dim M < \alpha$, and $\varphi(M) = 0$ for all M satisfying $\dim M > \alpha$.*

Proof. We may restrict ourselves to the space $\mathcal{C}_{0,1}$ of all compact subsets of $(0, 1)$. There exists a probability measure μ on $\mathcal{C}_{0,1}$ such that the function $\varphi(M) = \mu\{M_1 : M_1 \cap M = \emptyset\}$ satisfies two conditions: $\varphi(M) = 1$ for all M such that $\dim M < \alpha$, and $\varphi(M) < 1$ for all M such that $\dim M > \alpha$. That is a result of J. Hawkes, see [6, Th. 6], [10, Lemma 5.1]. By Lemma 6.12, $\varphi \in S$. By multiplicativity (of S), also $\varphi^n \in S$ for all n. The function $\lim_n \varphi^n$ satisfies the required conditions. $\square$

As a by-product we see that the Hausdorff dimension is a *Borel* function $C \to \mathbb{R}$. (To this end we use an additional limiting procedure, as in the proof of Theorem 6.14.)

Theorem 6.14. *Let* $((\Omega, \mathcal{F}, P), (\mathcal{F}_{s,t})_{s \le t})$ *be a continuous factorization, and* $\alpha \in (0,1)$ *a number. Then there exist sub-σ-fields* $\mathcal{E}_{\alpha-}, \mathcal{E}_{\alpha+}$ *of* $\mathcal{F}$ *such that for all* $f \in L_2(\Omega, \mathcal{F}, P)$,

(a) f *is measurable w.r.t.* $\mathcal{E}_{\alpha-}$ *if and only if* μ_f *is concentrated on the set of* $M \in C$ *such that* $\dim M < \alpha$;

(b) f *is measurable w.r.t.* $\mathcal{E}_{\alpha+}$ *if and only if* μ_f *is concentrated on the set of* $M \in C$ *such that* $\dim M \le \alpha$.

Proof. We choose $\alpha_k \to \alpha$, apply Lemma 6.13 for each k, consider the limit φ of corresponding functions φ_k, and use Lemma 5.9. The case $\alpha_k < \alpha$ leads to (a), the case $\alpha_k > \alpha$ leads to (b). □

A more general notion behind Theorems 5.10, 6.7 and 6.14 is an ideal. Recall that a subset I of C is called an ideal, if

$$M_1 \subset M_2, \ M_2 \in I \implies M_1 \in I,$$
$$M_1, M_2 \in I \implies (M_1 \cup M_2) \in I.$$

In particular, C_{finite} and $C_{\text{countable}}$ are ideals. For every finite or countable ordinal α, all $M \in C$ such that $M^{(\alpha)} = \emptyset$ are an ideal. For every $\alpha \in (0,1)$, all $M \in C$ such that $\dim M < \alpha$ are an ideal. The same holds for '$\dim M \le \alpha$'. All these ideals are shift-invariant:

$$M \in I \implies (M+t) \in I \quad \text{for all } t,$$
$$M + t = \{m + t : m \in M\},$$

but in general, an ideal need not be shift-invariant. Also, all ideals mentioned above are Borel subsets of C, except for $C_{\text{countable}}$; the latter is universally measurable, but not Borel. The following theorem is formulated for Borel ideals, but holds also for universally measurable ideals. Conditions 6.15 (a,b,c) parallel 3.16 (a,b,c), which means that sub-σ-fields $\mathcal{E}_{s,t}$ form a continuous factorization of the quotient probability space $(\Omega, \mathcal{F}, P)/\mathcal{E}$.

Theorem 6.15. *Let* $((\Omega, \mathcal{F}, P), (\mathcal{F}_{s,t})_{s \le t})$ *be a continuous factorization,* $I \subset C$ *a Borel ideal,* $\mathcal{E} \subset \mathcal{F}$ *a sub-σ-field, and for every* $f \in L_2(\Omega, \mathcal{F}, P)$, f *be* $\mathcal{E}$-*measurable if and only if* μ_f *is concentrated on* I. *Then sub-σ-fields* $\mathcal{E}_{s,t} = \mathcal{E} \cap \mathcal{F}_{s,t}$ *satisfy the conditions*

$$\mathcal{E}_{r,t} = \mathcal{E}_{r,s} \otimes \mathcal{E}_{s,t} \quad \text{whenever } r \le s \le t, \tag{a}$$

$$\bigcup_{\varepsilon > 0} \mathcal{E}_{s+\varepsilon, t-\varepsilon} \text{ generates } \mathcal{E}_{s,t} \text{ whenever } s < t, \tag{b}$$

$$\bigcup_{n=1}^{\infty} \mathcal{E}_{-n,n} \text{ generates } \mathcal{E}. \tag{c}$$

Proof. (a) We introduce Borel subsets $I_{s,t} = \{M \in I : M \subset (s,t)\}$ of $\mathcal{C}$ and the corresponding subspaces $H_{s,t} = H_{I_{s,t}}$ of $L_2(\Omega, \mathcal{F}, P)$. The equality $I_{r,t} = I_{r,s} \times I_{s,t}$ (treated according to (6.1)) follows easily from the fact that I is an ideal. Lemma 6.1 (or rather, its evident generalization) states that $H_{r,t} = H_{r,s} \otimes H_{s,t}$. On the other hand,

$$L_2(\mathcal{E}_{s,t}) = L_2(\mathcal{E} \cap \mathcal{F}_{s,t}) = L_2(\mathcal{E}) \cap L_2(\mathcal{F}_{s,t}) = H_I \cap H_{\mathcal{C}_{s,t}} = H_{I \cap \mathcal{C}_{s,t}} = H_{s,t}.$$

So, $L_2(\mathcal{E}_{r,t}) = L_2(\mathcal{E}_{r,s}) \otimes L_2(\mathcal{E}_{s,t})$, therefore $\mathcal{E}_{r,t} = \mathcal{E}_{r,s} \otimes \mathcal{E}_{s,t}$.

(c) $\cup_n I_{-n,n} = I$, therefore $\cup_n H_{I_{-n,n}}$ is dense in H_I; that is, $\cup_n L_2(\mathcal{E}_{-n,n})$ is dense in $L_2(\mathcal{E})$, therefore $\cup_n \mathcal{E}_{-n,n}$ generates $\mathcal{E}$.

(b): similarly to (c). $\qquad\square$

Remark 6.16. If the ideal I is shift-invariant and the given object is a noise (not only a factorization), then the sub-factorization $(\mathcal{E}_{s,t})$ becomes a sub-noise. In particular, every nonclassical noise has its classical (in other words, stable) sub-noise.

Question 6.17. Does every Borel ideal correspond to a sub-σ-field? (For an arbitrary continuous factorization, I mean. Though, the question is also open for noises and shift-invariant ideals.)

6.3 An Old Question of Jacob Feldman

Let $\big((\Omega, \mathcal{F}, P), (\mathcal{F}_{s,t})_{s \leq t}\big)$ be a continuous factorization. Sub-σ-fields $\mathcal{F}_E$ correspond to elementary sets $E \subset \mathbb{R}$ (recall Sect. 3.4) and satisfy

$$\mathcal{F}_{E_1 \cup E_2} = \mathcal{F}_{E_1} \otimes \mathcal{F}_{E_2} \quad \text{whenever } E_1 \cap E_2 = \emptyset. \tag{6.2}$$

It is natural to ask whether or not the map $E \mapsto \mathcal{F}_E$ can be extended to all Borel sets $E \subset \mathbb{R}$ in such a way that (6.2) is still satisfied and in addition,

$$\mathcal{F}_{E_n} \uparrow \mathcal{F}_E \quad \text{whenever } E_n \uparrow E. \tag{6.3}$$

The answer is positive if and only if the given continuous factorization is classical (Theorem 6.21 below, see also [18]), which solves a question of Feldman [4].

Note that (6.3) implies

$$\mathcal{F}_{E_n} \downarrow \mathcal{F}_E \quad \text{whenever } E_n \downarrow E. \tag{6.4}$$

Proof: Let $E_n \downarrow E$, then $\mathcal{F}_{\mathbb{R} \setminus E_n} \uparrow \mathcal{F}_{\mathbb{R} \setminus E}$ by (6.3), and so $\mathcal{F}_{\mathbb{R} \setminus E}$ is independent of $\cap \mathcal{F}_{E_n}$. If $\mathcal{F}_E$ is strictly less than $\cap \mathcal{F}_{E_n}$, then $\mathcal{F}_E \otimes \mathcal{F}_{\mathbb{R} \setminus E}$ is strictly less than $(\cap \mathcal{F}_{E_n}) \otimes \mathcal{F}_{\mathbb{R} \setminus E}$, which cannot happen, since $\mathcal{F}_E \otimes \mathcal{F}_{\mathbb{R} \setminus E} = \mathcal{F}$ by (6.2).

An extension satisfying (6.3), (6.4) is unique (if it exists) by the monotone class theorem. Therefore an extension (of $(\mathcal{F}_E)$) to the Borel σ-field) satisfying (6.2), (6.3) is unique (if it exists).

Lemma 6.18. *If the factorization is classical then an extension satisfying (6.2), (6.3) exists.*

Proof. By (slightly generalized) Theorem 6.2, for every elementary E, the σ-field $\mathcal{F}_E = \mathcal{F}_E^{\text{stable}}$ is generated by the corresponding portion $H_E^{(1)} = L_2(\mathcal{F}_E) \cap H^{(1)}$ of the first chaos $H^{(1)}$. The space $H_E^{(1)}$ corresponds (in the sense of Theorem 3.26) to the subset $\mathcal{M}_E^{(1)} \subset \mathcal{C}$ of all single-point subsets of E.

Given an arbitrary Borel set $E \subset \mathbb{R}$, we define the subset $\mathcal{M}_E^{(1)} \subset \mathcal{C}$ as above (that is, all single-point subsets of E), consider the corresponding subspace $H_E^{(1)} \subset H^{(1)}$, and introduce the sub-$\sigma$-field $\mathcal{F}_E \subset \mathcal{F}$ generated by $H_E^{(1)}$.

Given $f \in H^{(1)}$, we denote by f_E the orthogonal projection of f to $H_E^{(1)}$; here E is an arbitrary Borel set. If $E_n \uparrow E$ (or $E_n \downarrow E$) then $f_{E_n} \to f$ in L_2. If E is elementary then

$$\mathbb{E}\, e^{\mathrm{i}f} = \left(\mathbb{E}\, e^{\mathrm{i}f_E} \right) \left(\mathbb{E}\, e^{\mathrm{i}f_{\mathbb{R}\setminus E}} \right)$$

due to independence. The monotone class theorem extends the equality to all Borel sets E. We conclude that f_E and $f_{\mathbb{R}\setminus E}$ are independent. Therefore σ-fields $\mathcal{F}_E$ and $\mathcal{F}_{\mathbb{R}\setminus E}$ are independent for every Borel set E. Taking into account that $H_{E_1 \cup E_2}^{(1)} = H_{E_1}^{(1)} \oplus H_{E_2}^{(1)}$ whenever $E_1 \cap E_2 = \emptyset$ we get (6.2).

If $E_n \uparrow E$ then $H_{E_n}^{(1)} \uparrow H_E^{(1)}$, which ensures (6.3). $\square$

Condition (a) of the next lemma is evidently necessary for the extension to exist. In more topological language, for every open set $G \subset \mathbb{R}$ the corresponding σ-field $\mathcal{F}_G$ is naturally defined by approximation (of G by elementary sets) from within, while a closed set is approximated from the outside. The necessary condition, $\mathcal{F}_G \otimes \mathcal{F}_{\mathbb{R}\setminus G} = \mathcal{F}$, appears to be equivalent to the following (see 6.19(b)): the set $M \cap G$ is compact, for almost all $M \in \mathcal{C}$.

Lemma 6.19. *For all elementary sets $E_1 \subset E_2 \subset \dots$ the following two conditions are equivalent:*

(a) $\left(\bigvee_n \mathcal{F}_{E_n} \right) \otimes \left(\bigwedge_n \mathcal{F}_{\mathbb{R}\setminus E_n} \right) = \mathcal{F}$;

(b) *the set $\{ M \in \mathcal{C} : \forall n\ M \cap \left((\cup E_k) \setminus E_n \right) \neq \emptyset \}$ is negligible w.r.t. the spectral measure μ_f for every $f \in L_2(\Omega, \mathcal{F}, P)$.*

Proof. Denote $F_n = \mathbb{R} \setminus E_n$, $\mathcal{E}_n = \mathcal{F}_{E_n}$, $\mathcal{F}_n = \mathcal{F}_{\mathbb{R}\setminus E_n}$, $\mathcal{E}_\infty = \vee_n \mathcal{E}_n$, $\mathcal{F}_\infty = \wedge_n \mathcal{F}_n$. Clearly, $\mathcal{E}_\infty$ and $\mathcal{F}_\infty$ are independent, and (a) becomes $\mathcal{E}_\infty \vee \mathcal{F}_\infty = \mathcal{F}$. Denote also $\mathcal{M}_n = \{ M \in \mathcal{C} : M \subset E_n \}$, $\mathcal{N}_n = \{ M \in \mathcal{C} : M \subset F_n \}$, $\mathcal{M}_\infty = \cup_n \mathcal{M}_n = \{ M \in \mathcal{C} : \exists n\ M \subset E_n \}$, $\mathcal{N}_\infty = \cap_n \mathcal{N}_n = \{ M \in \mathcal{C} : M \subset \cap F_n \}$; then $H_{\mathcal{M}_n} = L_2(\mathcal{E}_n)$, $H_{\mathcal{N}_n} = L_2(\mathcal{F}_n)$. We have $\mathcal{M}_n \uparrow \mathcal{M}_\infty$ and $\mathcal{N}_n \downarrow \mathcal{N}_\infty$; therefore $L_2(\mathcal{E}_n) = H_{\mathcal{M}_n} \uparrow H_{\mathcal{M}_\infty}$ and $L_2(\mathcal{F}_n) = H_{\mathcal{N}_n} \downarrow H_{\mathcal{N}_\infty}$. On the other hand, $\mathcal{E}_n \uparrow \mathcal{E}_\infty$ and $\mathcal{F}_n \downarrow \mathcal{F}_\infty$; therefore $L_2(\mathcal{E}_n) \uparrow L_2(\mathcal{E}_\infty)$ and $L_2(\mathcal{F}_n) \downarrow L_2(\mathcal{F}_\infty)$. So,

$$H_{\mathcal{M}_\infty} = L_2(\mathcal{E}_\infty), \quad H_{\mathcal{N}_\infty} = L_2(\mathcal{F}_\infty).$$

Denote $\mathcal{M}_\infty \vee \mathcal{N}_\infty = \{M_1 \cup M_2 : M_1 \in \mathcal{M}_\infty, M_2 \in \mathcal{N}_\infty\}$; the same for $\mathcal{M}_1 \vee \mathcal{N}_\infty$ etc. We have $H_{\mathcal{M}_1 \vee \mathcal{N}_n} = H_{\mathcal{M}_1} \otimes H_{\mathcal{N}_n}$ and $\mathcal{M}_1 \vee \mathcal{N}_n \downarrow \mathcal{M}_1 \vee \mathcal{N}_\infty$; thus $H_{\mathcal{M}_1 \vee \mathcal{N}_\infty} = H_{\mathcal{M}_1} \otimes H_{\mathcal{N}_\infty}$ (note a relation to Lemma 6.1). Similarly, $H_{\mathcal{M}_n \vee \mathcal{N}_\infty} = H_{\mathcal{M}_n} \otimes H_{\mathcal{N}_\infty}$. However, $\mathcal{M}_n \vee \mathcal{N}_\infty \uparrow \mathcal{M}_\infty \vee \mathcal{N}_\infty$, and we get $H_{\mathcal{M}_\infty \vee \mathcal{N}_\infty} = H_{\mathcal{M}_\infty} \otimes H_{\mathcal{N}_\infty}$, that is,

$$H_{\mathcal{M}_\infty \vee \mathcal{N}_\infty} = L_2(\mathcal{E}_\infty) \otimes L_2(\mathcal{F}_\infty).$$

Now (a) becomes $H_{\mathcal{M}_\infty \vee \mathcal{N}_\infty} = H$, which means negligibility of the set $\mathcal{C} \setminus (\mathcal{M}_\infty \vee \mathcal{N}_\infty) = \{M : \forall n \; M \cap ((\cup E_k) \setminus E_n) \neq \emptyset\}$, that is, (b). $\qquad \square$

Every classical factorization satisfies 6.19(b), since a finite set M cannot intersect $(\cup E_k) \setminus E_n$ for all n.

Lemma 6.20. *If Condition 6.19(b) is satisfied for every (E_n) then the factorization is classical.*

Proof. Let the factorization be not classical. Then we can choose a sensitive $f \in L_2(\Omega, \mathcal{F}, P)$, $\|f\| = 1$. Assume for convenience that $f \in L_2(\mathcal{F}_{0,1})$, and consider the spectral measure μ_f; μ_f-almost all M are infinite subsets of $(0,1)$. We choose $p_1, p_2, \cdots \in (0,1)$ such that $\sum p_k \leq 1/3$ (say, $p_k = 2^{-k}/3$). Integer parameters $n_1 < n_2 < \ldots$ will be chosen later. We introduce independent random elementary sets $B_1, B_2, \cdots \subset [0,1]$ as follows:

$$\mathbb{P}\left\{ B_k = \left(\frac{l_1-1}{n_k}, \frac{l_1}{n_k}\right) \cup \cdots \cup \left(\frac{l_m-1}{n_k}, \frac{l_m}{n_k}\right) \right\} = p_k^m (1-p_k)^{n_k - m}$$

whenever $1 \leq l_1 < \cdots < l_m \leq n_k$, $m \in \{0, \ldots, n_k\}$. That is, we have a two-parameter family of independent events, $\left(\frac{l-1}{n_k}, \frac{l}{n_k}\right) \subset B_k$, where $l \in \{1, \ldots, n_k\}$, $k \in \{1, 2, \ldots\}$. The probability of such an event is equal to p_k. We define $E_k = B_1 \cup \cdots \cup B_k$; thus $E_1 \subset E_2 \subset \ldots$ is a (random) increasing sequence of elementary subsets of $[0,1]$.

We treat M as a random compact subset of $(0,1)$, distributed μ_f and independent of $B_1, B_2, \ldots$ Let $\tilde{P}$ be the corresponding probability measure (in fact, product measure) on the space $\tilde{\Omega}$ of sequences (of sets) $(M, B_1, B_2, \ldots)$. For each $k = 0, 1, 2, \ldots$ we define an event A_k, that is, a measurable subset of $\tilde{\Omega}$, by the following condition on $(M, B_1, B_2, \ldots)$:

$$M \setminus E_k \text{ is infinite and does not intersect } B_{k+1};$$

of course, $E_0 = \emptyset$.

We can choose $n_1, n_2, \ldots$ such that $\sum_k \tilde{P}(A_k) \leq 1/3$. Proof: $\tilde{P}(A_k)$ is a function of $n_1, \ldots, n_k, n_{k+1}$ that converges to 0 when $n_{k+1} \to \infty$ (while $n_1, \ldots, n_k$ are fixed).

The probability of the event

$$M \setminus E_k \text{ is infinite for all } k$$

is no less than $1 - \sum p_k \geq 2/3$. Proof: Each M has a limit point (at least one), and the point is covered by (the closure of) $B_1 \cup B_2 \cup \ldots$ with probability $\leq \sum p_k$.

So, there is a positive probability $(\geq 1/3)$ to such an event:

for each k, the set $M \setminus E_k$ is infinite and intersects B_{k+1}.

However, the conditional probability, given $B_1, B_2, \ldots$ (but not M) of the event

for each k, the set $M \setminus E_k$ intersects B_{k+1}

must vanish according to Condition 6.19(b).   □

Theorem 6.21. *A continuous factorization is classical if and only if the map $E \mapsto \mathcal{F}_E$ can be extended from the algebra of elementary sets to the Borel σ-field, satisfying (6.2) and (6.3).*

Proof. If the factorization is classical then the extension exists by Lemma 6.18. Let the extension exist; then 6.19(a) is satisfied for all (E_k), therefore 6.19(b) is also satisfied, and the factorization is classical by Lemma 6.20.   □

6.4 Black Noise

Definition 6.22. *A noise is* black, *if its stable σ-field $\mathcal{F}_{\text{stable}}$ is degenerate. In other words: its first chaos contains only 0.*

Why 'black'? Well, the white noise is called 'white' since its spectral density is constant. It excites harmonic oscillators of all frequencies to the same extent. For a black noise, however, the response of any linear sensor is zero!

What could be a physically reasonable nonlinear sensor able to sense a black noise? Maybe a fluid could do it, which is hinted at by the following words of Shnirelman [13, p. 1263] about the paradoxical motion of an ideal incompressible fluid: '... very strong external forces are present, but they are infinitely fast oscillating in space and therefore are indistinguishable from zero in the sense of distributions. The smooth test functions are not "sensitive" enough to "feel" these forces.'

The very idea of black noises, nonclassical factorizations, etc. was suggested to me by Anatoly Vershik in 1994.

Lemma 6.23. *Let $\big((\Omega, \mathcal{F}, P), (\mathcal{F}_{s,t})_{s \leq t}\big)$ be a continuous factorization, $a < b$, $\mathcal{M}$ a Borel subset of $\mathcal{C}_{a,b} = \{M \in \mathcal{C} : M \subset (a, b)\}$, and $\tilde{\mathcal{M}} = \{M \in \mathcal{C} : M \cap (a, b) \in \mathcal{M}\}$. If $\mu_f(\mathcal{M}) = 0$ for all $f \in L_2(\Omega, \mathcal{F}, P)$ then $\mu_f(\tilde{\mathcal{M}}) = 0$ for all $f \in L_2(\Omega, \mathcal{F}, P)$.*

Proof. I prove it for $(a, b) = (0, \infty)$, leaving the general case to the reader. We have $\mathcal{C} = \mathcal{C}_{-\infty,0} \times \mathcal{C}_{0,\infty}$, $\mathcal{M} \subset \mathcal{C}_{0,\infty}$ and $\tilde{\mathcal{M}} = \mathcal{C}_{-\infty,0} \times \mathcal{M}$ (in the sense of (6.1)). By Lemma 6.1, $H_{\tilde{\mathcal{M}}} = H_{\mathcal{C}_{-\infty,0} \times \mathcal{M}} = H_{\mathcal{C}_{-\infty,0}} \otimes H_{\mathcal{M}}$. By (3.12), the space $H_{\mathcal{M}}$ is trivial (that is, $\{0\}$). Therefore $H_{\tilde{\mathcal{M}}}$ is also trivial; it remains to use (3.12) again.   □

Recall that a compact set M is called perfect, if it has no isolated points. (The empty set is also perfect.) The set $\mathcal{C}_{\text{perfect}}$ of all perfect compact subsets of $\mathbb{R}$ is a Borel set in $\mathcal{C}$, see [7, proof of Th. 27.5].

Theorem 6.24. *For every continuous factorization $((\Omega, \mathcal{F}, P), (\mathcal{F}_{s,t})_{s \leq t})$ the following two conditions are equivalent:*

(a) *the first chaos space is trivial (contains only 0);*

(b) *for every $f \in L_2(\Omega, \mathcal{F}, P)$ the spectral measure μ_f is concentrated on $\mathcal{C}_{\text{perfect}}$.*

Proof. (b) implies (a) evidently (a single-point set cannot be perfect). Assume (a). Applying Lemma 6.23 to the set $\mathcal{M}$ of all single-point subsets of (a, b) we see that μ_f-almost all $M \in \mathcal{C}$ are such that $M \cap (a, b)$ is not a single-point set, for all rational $a < b$. It means that M is perfect.   □

So, a noise is black if and only if spectral measures are concentrated on (the set of all) perfect sets.

Existence of black noises was proven first by Tsirelson and Vershik [20, Sect. 5]. A simpler and more natural example is described in the next section. Another example is found by Watanabe [26].

If all spectral sets are finite or countable (as in Examples 6.9, 6.10), such a noise cannot contain a black sub-noise.

Question 6.25. If a noise contains no black sub-noise, does it follow that all spectral sets are at most countable?

Perfect sets may be classified, say, by Hausdorff dimension. For any $\alpha \in (0, 1)$, sets $M \in \mathcal{C}$ of Hausdorff dimension $\leq \alpha$ are a shift invariant ideal, corresponding to a sub-noise. Also, all $M \in \mathcal{C}$ of Hausdorff dimension α correspond to a 'chaos subspace number α'. A continuum of such chaos subspaces (not in a single noise, of course) could occur, describing different 'levels of sensitivity'. For now, however, I know of perfect spectral sets of Hausdorff dimension $1/2$ only.

Question 6.26. Can a noise have perfect spectral sets of Hausdorff dimension other than $1/2$? (See also the end of Sect. 8.3.)

Question 6.27. Can a black noise emerge as the refinement of a dyadic coarse factorization satisfying (3.13)?

The following results (especially Corollary 6.35) may be treated as continuous-time counterparts of Theorem 5.3 (of Benjamini, Kalai and Schramm). Given a continuous factorization $((\Omega, \mathcal{F}, P), (\mathcal{F}_{s,t})_{s \le t})$ and a function $f \in L_2(\Omega, \mathcal{F}, P)$, we define

$$\mathbf{H}(f) = \limsup_{\{t_1, \ldots, t_n\} \uparrow} \sum_{k=1}^{n+1} \left(\mathbb{E} \sqrt{\mathrm{Var}\left(f \mid \mathcal{F}_{\mathbb{R} \setminus (t_{k-1}, t_k)} \right)} \right)^2 ;$$

here $t_0 = -\infty$, $t_{n+1} = +\infty$, and the 'lim sup' is taken over all finite sets $L = \{t_1, \ldots, t_n\} \subset \mathbb{R}$, $t_1 < \cdots < t_n$, ordered by inclusion. That is, 'for every ε there exists L_ε such that for all $L \supset L_\varepsilon \ldots$' and so on. We also introduce

$$\mathbf{H}_1(f) = \lim_{\{t_1, \ldots, t_n\} \uparrow} \sum_{k=1}^{n+1} \mathrm{Var}(\mathbb{E}\left(f \mid \mathcal{F}_{t_{k-1}, t_k} \right)) .$$

This time we may write 'lim' (or 'inf') instead of 'lim sup' due to monotonicity (w.r.t. inclusion); the more $L = \{t_1, \ldots, t_n\}$ the less the sum.

Lemma 6.28. $\sqrt{\mathrm{Var}(\mathbb{E}\left(f \mid \mathcal{F}_{s,t} \right))} \le \mathbb{E} \sqrt{\mathrm{Var}\left(f \mid \mathcal{F}_{\mathbb{R} \setminus (s,t)} \right)}$ *for all* $f \in L_2(\Omega, \mathcal{F}, P)$ *and* $s < t$.

Proof. The space $L_2(\Omega, \mathcal{F}, P) = L_2(\mathcal{F}) = L_2(\mathcal{F}_{s,t} \otimes \mathcal{F}_{\mathbb{R} \setminus (s,t)}) = L_2(\mathcal{F}_{s,t}) \otimes L_2(\mathcal{F}_{\mathbb{R} \setminus (s,t)})$ may also be thought of as the space $L_2(\mathcal{F}_{\mathbb{R} \setminus (s,t)}, L_2(\mathcal{F}_{s,t}))$ consisting of $\mathcal{F}_{\mathbb{R} \setminus (s,t)}$-measurable square integrable vector-functions, taking on values in $L_2(\mathcal{F}_{s,t})$. We consider the element $\tilde{f} \in L_2(\mathcal{F}_{\mathbb{R} \setminus (s,t)}, L_2(\mathcal{F}_{s,t}))$ corresponding to $f \in L_2(\mathcal{F})$ (according to the canonical isomorphism of these two spaces). The mean value of the vector-function is $\mathbb{E}\, \tilde{f} = \mathbb{E}\left(f \mid \mathcal{F}_{s,t} \right)$ (these two '$\mathbb{E}$' act on different spaces). Convexity of the seminorm $\sqrt{\mathrm{Var}(\cdot)}$ on $L_2(\mathcal{F}_{s,t})$ gives $\sqrt{\mathrm{Var}(\mathbb{E}\, \tilde{f})} \le \mathbb{E} \sqrt{\mathrm{Var}(\tilde{f})}$, where $\mathrm{Var}(\tilde{f})$ means the pointwise variance (each value of $\tilde{f}$ is a random variable; the latter has its variance), basically the same as $\mathrm{Var}\left(f \mid \mathcal{F}_{\mathbb{R} \setminus (s,t)} \right)$. $\square$

Corollary 6.29. $\mathbf{H}_1(f) \le \mathbf{H}(f)$.

Lemma 6.30. $\mathbf{H}_1(f) = \|Q_1 f\|$ *for all* $f \in L_2(\Omega, \mathcal{F}, P)$; *here* Q_1 *is the orthogonal projection onto the first chaos.*

Proof. Follows immediately from Theorem 6.3. $\square$

Corollary 6.31. *Every* $f \in L_2(\Omega, \mathcal{F}, P)$ *such that* $\mathbf{H}(f) = 0$ *is orthogonal to the first chaos.*

Corollary 6.32. *If a noise is such that* $\mathbf{H}(f) = 0$ *for all* $f \in L_2(\Omega, \mathcal{F}, P)$, *then the noise is black.*

Lemma 6.33. *Let $g \in L_2(\mathcal{F})$, $h \in L_\infty(\mathcal{F}_{0,\infty})$, and $f = \mathbb{E}\left(gh \,\middle|\, \mathcal{F}_{-\infty,0} \right)$. Then $\mathbf{H}(f) \leq \|h\|_\infty^2 \mathbf{H}(g)$.*

Proof. It is sufficient to prove the inequality for the influence, $\mathbb{E}\sqrt{\operatorname{Var}\left(f \,\middle|\, \mathcal{F}_{\mathbb{R}\setminus(s,t)} \right)} \leq \|h\|_\infty \mathbb{E}\sqrt{\operatorname{Var}\left(g \,\middle|\, \mathcal{F}_{\mathbb{R}\setminus(s,t)} \right)}$ for any $(s,t) \subset (-\infty, 0)$. Similarly to the proof of Lemma 6.28, we consider $\tilde{g} \in L_2(\mathcal{F}_{0,\infty}, L_2(\mathcal{F}_{-\infty,0}))$ corresponding to $g \in L_2(\mathcal{F}_{-\infty,0} \otimes \mathcal{F}_{0,\infty})$. We have $\tilde{g}h \in L_2(\mathcal{F}_{0,\infty}, L_2(\mathcal{F}_{-\infty,0}))$, $\mathbb{E}(\tilde{g}h) = f$. Convexity of the seminorm $\mathbb{E}\sqrt{\operatorname{Var}\left(\cdot \,\middle|\, \mathcal{F}_{(-\infty,0)\setminus(s,t)} \right)}$ on $L_2(\mathcal{F}_{-\infty,0})$ gives $\mathbb{E}\sqrt{\operatorname{Var}\left(f \,\middle|\, \mathcal{F}_{(-\infty,0)\setminus(s,t)} \right)} \leq \mathbb{E}\mathbb{E}\sqrt{\operatorname{Var}\left(\tilde{g}h \,\middle|\, \mathcal{F}_{(-\infty,0)\setminus(s,t)} \right)}$, where 'Var' and the internal '$\mathbb{E}$' act on $L_2(\mathcal{F}_{-\infty,0})$, while the outer '$\mathbb{E}$' acts on $L_2(\mathcal{F}_{0,\infty})$. The right-hand side is equal to $\mathbb{E}\left(|h|\mathbb{E}\sqrt{\operatorname{Var}\left(\tilde{g} \,\middle|\, \mathcal{F}_{(-\infty,0)\setminus(s,t)} \right)} \right)$ and so, cannot exceed $\|h\|_\infty \mathbb{E}\mathbb{E}\sqrt{\operatorname{Var}\left(\tilde{g} \,\middle|\, \mathcal{F}_{(-\infty,0)\setminus(s,t)} \right)} = \|h\|_\infty \mathbb{E}\sqrt{\operatorname{Var}\left(g \,\middle|\, \mathcal{F}_{\mathbb{R}\setminus(s,t)} \right)}$. □

Lemma 6.34. *If $f \in L_2(\Omega, \mathcal{F}, P)$ is such that $\mathbf{H}(f) = 0$, then μ_f is concentrated on $\mathcal{C}_{\mathrm{perfect}}$.*

Proof. Similarly to the proof of Theorem 6.24, it is sufficient to prove, for every $(a,b) \subset \mathbb{R}$, that μ_f-almost all $M \in \mathcal{C}$ are such that $M \cap (a,b)$ is not a single-point set. Lemma 6.1 shows that the subspace corresponding to $\{M \in \mathcal{C} : |M \cap (a,b)| = 1\}$ is $H_{-\infty,a} \otimes H_{a,b}^{(1)} \otimes H_{b,\infty}$, where $H_{a,b}^{(1)}$ is the first chaos intersected with $H_{a,b}$. We have to prove that f is orthogonal to $H_{-\infty,a} \otimes H_{a,b}^{(1)} \otimes H_{b,\infty}$, that is, to gh for every $g \in H_{a,b}^{(1)}$, $h \in H_{-\infty,a} \otimes H_{b,\infty} = L_2(\mathcal{F}_{\mathbb{R}\setminus(a,b)})$, and we may assume that $h \in L_\infty(\mathcal{F}_{\mathbb{R}\setminus(a,b)})$.

We have $\mathbb{E}(fgh) = \mathbb{E}\left(g\mathbb{E}\left(fh \,\middle|\, \mathcal{F}_{a,b} \right) \right)$. Lemma 6.33 (slightly generalized) shows that $\mathbf{H}\left(\mathbb{E}\left(fh \,\middle|\, \mathcal{F}_{a,b} \right) \right) \leq \|h\|_\infty^2 \mathbf{H}(f)$. Thus, $\mathbf{H}\left(\mathbb{E}\left(fh \,\middle|\, \mathcal{F}_{a,b} \right) \right) = 0$; by Corollary 6.31, $\mathbb{E}\left(g\mathbb{E}\left(fh \,\middle|\, \mathcal{F}_{a,b} \right) \right) = 0$. □

Corollary 6.35. *Let $\left((\Omega, \mathcal{F}, P), (\mathcal{F}_{s,t})_{s \leq t} \right)$ be a continuous factorization. If $f \in L_2(\Omega, \mathcal{F}, P)$ satisfies $\mathbf{H}(f) = 0$ and $\mathbb{E}\, f = 0$, then f is sensitive.*

Here are counterparts of Lemma 5.7 and Theorem 5.10 inspired by the work [9] of Le Jan and Raimond.

Lemma 6.36. *Let $f \in L_2(\Omega, \mathcal{F}, P)$, and $g = \eta \circ f$ where $\eta : \mathbb{R} \to \mathbb{R}$ satisfies $|\eta(x) - \eta(y)| \leq |x - y|$ for all $x, y \in \mathbb{R}$. Then*

$$\mathbf{H}(g) \leq \mathbf{H}(f).$$

Proof. It is sufficient to prove the inequality for the influence,

$$\mathbb{E}\sqrt{\operatorname{Var}\left(g \,\middle|\, \mathcal{F}_{\mathbb{R}\setminus(s,t)} \right)} \leq \mathbb{E}\sqrt{\operatorname{Var}\left(f \,\middle|\, \mathcal{F}_{\mathbb{R}\setminus(s,t)} \right)},$$

or a stronger inequality $\mathrm{Var}\left(g \mid \mathcal{F}_E\right) \leq \mathrm{Var}\left(f \mid \mathcal{F}_E\right)$ a.s., for an arbitrary elementary set E. It is a conditional counterpart of the inequality $\mathrm{Var}(\eta \circ X) \leq \mathrm{Var}(X)$ for any random variable X. A proof of the latter: $\mathrm{Var}(\eta \circ X) = \frac{1}{2}\mathbb{E}\left(\eta \circ X_1 - \eta \circ X_2\right)^2 \leq \frac{1}{2}\mathbb{E}\left(X_1 - X_2\right)^2 = \mathrm{Var}(X)$, where X_1, X_2 are independent copies of X. $\qquad\square$

Theorem 6.37. *For every continuous factorization* $\left((\Omega, \mathcal{F}, P), (\mathcal{F}_{s,t})_{s \leq t}\right)$ *there exists a sub-σ-field* $\mathcal{F}_{\mathrm{jetblack}}$ *of* $\mathcal{F}$ *such that* $L_2(\Omega, \mathcal{F}_{\mathrm{jetblack}}, P)$ *is the closure (in* $L_2(\Omega, \mathcal{F}, P)$*) of* $\{f \in L_2(\Omega, \mathcal{F}, P) : \mathbf{H}(f) = 0\}$.

Proof. The set $\{f : \mathbf{H}(f) = 0\}$ is closed under linear operations, and also under the nonlinear operation $f \mapsto |f|$, therefore its closure is of the form $L_2(\mathcal{F}_{\mathrm{jetblack}})$. $\qquad\square$

Corollary 6.38. $L_2(\mathcal{F}_{\mathrm{jetblack}}) \subset H_{C_{\mathrm{perfect}}}$.

Question 6.39. Whether $\mathcal{F}_{\mathrm{jetblack}}$ is nontrivial for every black noise, or not?

7 Example: The Brownian Web as a Black Noise

7.1 Convolution Semigroup of the Brownian Web

A one-dimensional array of random signs can produce some classical and non-classical noises in the scaling limit, but I still do not know whether it can produce a black noise, or not (see Question 6.27).

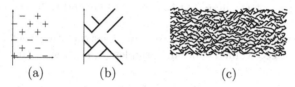

(a)   (b)   (c)

This is why I turn to a two-dimensional array of random signs (a). It produces a system of coalescing random walks (b) that converges to the so-called *Brownian web* (c), consisting of infinitely many coalescing Brownian motions (independent before coalescence).

The Brownian web was investigated by Arratia, Toth, Werner, Soucaliuc, and recently by Fontes, Isopi, Newman and Ravishankar [5] (other references may be found therein). The scaling limit may be interpreted in several ways, depending on the choice of 'observables', and may involve delicate points, because of complicated topological properties of the Brownian web as a random geometric configuration on the plane. However, we avoid these delicate points by treating the Brownian web as a stochastic flow in the sense of Sect. 4, that is, a two-parameter family of random variables in a semigroup.

In order to keep finite everything that can be kept finite, we consider Brownian motions in the circle $\mathbb{T} = \mathbb{R}/\mathbb{Z}$ rather than the line $\mathbb{R}$.

It is well-known that a countable dense set of coalescing 'particles', given at the initial instant, becomes finite, due to coalescence, after any positive time. Moreover, the finite number is of finite expectation. Thus, for any given $t > 0$, the Brownian web on the time interval $(0, t)$ gives us a random map $\mathbb{T} \to \mathbb{T}$ of the following elementary form (a step function):

$$f^{y_1,\ldots,y_n}_{x_1,\ldots,x_n} : \mathbb{T} \to \mathbb{T},$$

$$x_1 < \cdots < x_n < x_1, \; y_1 < \cdots < y_n < y_1 \text{ (cyclically)},$$

$$f^{y_1,\ldots,y_n}_{x_1,\ldots,x_n}(x) = y_{k+1} \text{ for } x \in (x_k, x_{k+1}].$$

Of course, n is random, as well as $x_1, \ldots, x_n$ and $y_1, \ldots, y_n$. The value at x_k does not matter; we let it be y_k for convenience, but it could equally well be y_{k+1}, or remain undefined. Points $x_1, \ldots, x_n$ will be called left critical points of the map, while $y_1, \ldots, y_n$ are right critical points.

We introduce the set G_∞ consisting of all step functions $\mathbb{T} \to \mathbb{T}$ and, in addition, the identity function. If $f, g \in G_\infty$ then their composition fg belongs to G_∞; thus G_∞ is a semigroup. It consists of pieces of dimensions $2, 4, 6, \ldots$ and the identity. Similarly to G_3 (recall (4.16)), G_∞ is not a topological semigroup, since the composition is discontinuous.

The distribution of the random map is a probability measure μ_t on G_∞. These maps form a convolution semigroup, $\mu_s * \mu_t = \mu_{s+t}$. Similarly to Sect. 4.5, discontinuity of composition does not harm, since the composition is continuous almost everywhere (w.r.t. $\mu_s \otimes \mu_t$). Left and right critical points do not meet.[32]

Having the convolution semigroup, we can construct the stochastic flow, that is, a family of G_∞-valued random variables $(\xi_{s,t})_{s \leq t}$ such that

$$\xi_{s,t} \sim \mu_{t-s},$$
$$\xi_{r,s}\xi_{s,t} = \xi_{r,t} \quad \text{a.s.}$$

whenever $-\infty < r < s < t < \infty$, and

$$\xi_{t_1,t_2}, \ldots, \xi_{t_{n-1},t_n} \quad \text{are independent}$$

whenever $-\infty < t_1 < \cdots < t_n < \infty$.

Indeed, for each i, we can take independent $\xi_{k/i,(k+1)/i} : \Omega[i] \to G_\infty$ for $k \in \mathbb{Z}$ according to the discrete model, and define $\xi_{k/i,l/i} = \xi_{k/i,(k+1)/i} \cdots \xi_{(l-1)/i,l/i}$. For any two coarse instants $s \leq t$, the distribution of $\xi_{s[i],t[i]}$ converges weakly (for $i \to \infty$) to $\mu_{t[\infty]-s[\infty]}$. The refinement gives us

$$\xi_{s,t} : \Omega \to G_\infty, \qquad \xi_{s,t} = f^{y_1(s,t),\ldots,y_{n(s,t)}(s,t)}_{x_1(s,t),\ldots,x_{n(s,t)}(s,t)};$$

$x_k(\cdot,\cdot)$ and $y_k(\cdot,\cdot)$ are continuous a.s. Also,

$$\mathbb{E}\, n(s,t) < \infty. \tag{7.1}$$

We consider the sub-σ-field $\mathcal{F}_{s,t}$ generated by all $\xi_{u,v}$ for $(u,v) \subset (s,t)$ and get a continuous factorization. Time shifts are evidently introduced, and so, we get a noise — the *noise of coalescence*.

7.2 Some General Arguments

Probably we could use $\mathbf{H}$ and Theorem 6.37 in order to prove that the noise of coalescence is black (see also [9]). However, I choose another way (via $\mathbf{H}_1$ rather than $\mathbf{H}$).

Random variables of the form $\varphi(\xi_{s,t})$ for arbitrary $s < t$ and arbitrary bounded Borel function $\varphi : G_\infty \to \mathbb{R}$ generate the whole σ-field $\mathcal{F}$. Products

---

[32] They meet with probability 0, as long as s and t are fixed. Otherwise, delicate points are involved...

of the form $\varphi_1(\xi_{t_0,t_1})\ldots\varphi_n(\xi_{t_{n-1},t_n})$ for $t_0 < \cdots < t_n$ span L_2 (as a closed subspace); however, we cannot expect that *linear* combinations of such $\varphi(\xi_{s,t})$ are dense in L_2.

Denote by Q_1 the orthogonal projection of $L_2(\Omega, \mathcal{F}, P)$ onto the first chaos.

Lemma 7.1. *Linear combinations of all $Q_1\varphi(\xi_{s,t})$ are dense in the first chaos.*

Proof: Follows easily from the next (quite general) result, or rather, its evident generalization to n factors.

Lemma 7.2. *Let $r \leq s \leq t$, $X \in L_2(\mathcal{F}_{r,s})$, $Y \in L_2(\mathcal{F}_{s,t})$. Then $Q_1(XY) = Q_1(X)\mathbb{E}(Y) + \mathbb{E}(X)Q_1(Y)$.*

Proof. In terms of operators R_φ given by Lemma 5.4 we have $Q_1(XY) = R_{\varphi_{r,t}}(XY)$, where $\varphi_{r,t} : \mathcal{C}_{r,t} \to \mathbb{R}$ is the indicator of $\{M \in \mathcal{C} : |M \cap (r,t)| = 1\}$. Similarly, $Q_1(X) = R_{\varphi_{r,s}}(X)$, and $\mathbb{E}(X) = R_{\psi_{r,s}}(X)$, where $\psi_{r,s}$ is the indicator of $\{M \in \mathcal{C} : |M \cap (r,s)| = 0\}$. However, $\varphi_{r,t} = \varphi_{r,s}\psi_{s,t} + \psi_{r,s}\varphi_{s,t}$ almost everywhere on $\mathcal{C}_{r,t}$ (w.r.t. every spectral measure). $\qquad\square$

In order to prove that the noise (of coalescence) is black, it suffices to prove that $Q\varphi(\xi_{s,t}) = 0$ for all s, t, φ. We'll prove that $Q\varphi(\xi_{0,1}) = 0$; the general case is similar. According to Lemma 6.30 we have to prove that $\mathbf{H}_1(\varphi(\xi_{0,1})) = 0$. Assuming that $\mathbb{E}\,\varphi(\xi_{0,1}) = 0$ we will check the sufficient condition:

$$\|\mathbb{E}\left(\varphi(\xi_{0,1})\,\big|\,\mathcal{F}_{t-\varepsilon,t}\right)\| = o(\sqrt{\varepsilon}) \quad \text{for } \varepsilon \to 0\,,$$

uniformly in t. When doing so, we may assume that t is bounded away from 0 and 1. Indeed, $\|\mathbb{E}\left(\varphi(\xi_{0,1})\,\big|\,\mathcal{F}_{t,1}\right)\| \to 0$ for $t \to 1-$, due to continuity of the factorization (recall Def. 3.16(b)).

Lemma 7.3. $\mathbb{E}\left(\varphi(\xi_{0,1})\,\big|\,\mathcal{F}_{t-\varepsilon,t}\right) = \mathbb{E}\left(\varphi(\xi_{0,1})\,\big|\,\xi_{t-\varepsilon,t}\right).$

The proof is left to the reader; a hint:

$$\mathbb{E}\left(\varphi(\xi_{t_1,t_5})\,\big|\,\xi_{t_2,t_3},\xi_{t_3,t_4}\right) = \iint \varphi(\xi_{12}\xi_{23}\xi_{34}\xi_{45})\,\mathrm{d}\mu_{t_2-t_1}(\xi_{12})\mathrm{d}\mu_{t_5-t_4}(\xi_{45})$$

$$= \mathbb{E}\left(\varphi(\xi_{t_1,t_5})\,\big|\,\xi_{t_2,t_4}\right).$$

7.3 The Key Argument

Similarly to Example 6.5, we consider $X = \varphi(\xi_{0,1}) = \varphi(\xi_{0,t-\varepsilon}\xi_{t-\varepsilon,t}\xi_{t,1})$, $\mathbb{E}\,X = 0$, $|X| \leq 1$ a.s. We have to prove that $\|\mathbb{E}\left(X\,\big|\,\xi_{t-\varepsilon,t}\right)\| = o(\sqrt{\varepsilon})$ for $\varepsilon \to 0$, uniformly in t, when t is bounded away from 0 and 1. Clearly,

$$\mathbb{E}\left(X\,\big|\,\xi_{t-\varepsilon,t}\right) = \iint \varphi(fgh)\,\mathrm{d}\mu_{t-\varepsilon}(f)\mathrm{d}\mu_{1-t}(h)\,,$$

where $g = \xi_{t-\varepsilon,t}$.

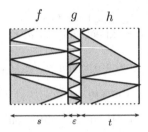

We choose $\gamma \in \left(\frac{1}{3}, \frac{1}{2}\right)$ and divide the strip $(t - \varepsilon, t) \times \mathbb{T}$ into $\sim \varepsilon^{-\gamma}$ 'cells' $(t - \varepsilon, t) \times (z_k, z_{k+1})$ of height $z_{k+1} - z_k \sim \varepsilon^\gamma$.

We want to think of g as consisting of independent cells. Probably it can be done in continuous time, but we have no such technique for now. Instead, we retreat to the discrete-time model. The needed inequality for continuous time results in the scaling limit $i \to \infty$ provided that in discrete time our estimations are uniform in i (for i large enough).

So, random signs that produce g are divided into cells. Cells are independent and, taken together, they determine g uniquely.

However, a path may cross many cells. This is rather improbable, since $\gamma < 1/2$, but it may happen. We enforce locality by a forgery! Namely, if the path starting at the middle of a cell reaches the bottom or the top edge of the cell, we replace the whole cell with some other cell (it may be chosen once and for all) where it does not happen.

Now cells are 'local'; a path cannot cross more than two cells, but of course, the stochastic flow is changed. Namely, g is changed with an exponentially small (for $\varepsilon \to 0$) probability, which changes $\mathbb{E}\left(X \mid \xi_{t-\varepsilon, t}\right)$ by $o(\sqrt{\varepsilon})$ (much less, in fact). Still, cells are independent.

Does a cell (of g) influence the composition, fgh? It depends on f and h. If the left edge $\{t - \varepsilon\} \times [z_k, z_{k+1}]$ of the cell contains no right critical point of f, the cell can influence, since a path starting in an adjacent cell can cross the boundary between cells. However, if the enlarged left edge $\{t - \varepsilon\} \times [z_k - \varepsilon^\gamma, z_{k+1} + \varepsilon^\gamma]$ contains no right critical point of f (in which case we say 'the cell is blocked by f'), then the cell cannot influence, because of the enforced locality. Similarly, if the enlarged right edge $\{t\} \times [z_k - \varepsilon^\gamma, z_{k+1} + \varepsilon^\gamma]$ contains no left critical point of h (in which case we say 'the cell is blocked by h'), the cell cannot influence.

The probability of being not blocked by f is the same for all cells, since the distribution of f is invariant under rotations of $\mathbb{T}$ (discretized as needed). The sum of these probabilities does not exceed $3\mathbb{E}\, n(0, t - \varepsilon)$ (recall (7.1)), which is $O(1)$ when $\varepsilon \to 0$. (Here we need t to be bounded away from 0.) Thus,

$$\mathbb{P}\left(\text{a given cell is not blocked by } f\right) = O(\varepsilon^\gamma);$$
$$\mathbb{P}\left(\text{a given cell is not blocked by } h\right) = O(\varepsilon^\gamma);$$
$$\mathbb{P}\left(\text{a given cell is not blocked}\right) = O(\varepsilon^{2\gamma});$$
$$\mathbb{P}\left(\text{at least one cell is not blocked}\right) = O(\varepsilon^\gamma).$$

In the latter case we may say that g is not blocked (by f, h).

Denote by A the event "g is not blocked by f, h" (it is determined by f and h, not g); $\mathbb{P}\left(A\right) = O(\varepsilon^\gamma)$. Taking into account that

$$X = X - \mathbb{E}X = \left(X \cdot 1_A - \mathbb{E}\left(X \cdot 1_A\right)\right) + \left(X \cdot (1 - 1_A) - \mathbb{E}\left(X \cdot (1 - 1_A)\right)\right),$$
$$\mathbb{E}\left(X \cdot (1 - 1_A)\,\middle|\, g\right) = \mathbb{E}\left(X \cdot (1 - 1_A)\right),$$
$$\mathbb{E}\left(X\,\middle|\, g\right) = \mathbb{E}\left(X \cdot 1_A\,\middle|\, g\right) - \mathbb{E}\left(X \cdot 1_A\right),$$

we have to prove that $\left\|\mathbb{E}\left(X \cdot 1_A\,\middle|\, g\right) - \mathbb{E}\left(X \cdot 1_A\right)\right\| = o(\sqrt{\varepsilon})$. Note that it does not result from the trivial estimation $\|X \cdot 1_A\| \le \|1_A\| = \sqrt{\mathbb{P}\left(A\right)} = O(\varepsilon^{\gamma/2})$, $\gamma \in \left(\frac{1}{3}, \frac{1}{2}\right)$. Note also that, when g influences X, its influence is usually not small (irrespective of ε) because of the stepwise nature of f and h.

We express the norm in terms of covariance,

$$\left\|\mathbb{E}\left(X \cdot 1_A\,\middle|\, g\right) - \mathbb{E}\left(X \cdot 1_A\right)\right\| = \sup_\psi \mathrm{Cov}\left(X \cdot 1_A, \psi(g)\right),$$

where the supremum is taken over all Borel functions $\psi : G_\infty \to \mathbb{R}$ such that $\mathrm{Var}\left(\psi(g)\right) \le 1$. In terms of the correlation coefficient

$$\mathrm{Corr}\left(X \cdot 1_A, \psi(g)\right) = \frac{\mathrm{Cov}\left(X \cdot 1_A, \psi(g)\right)}{\sqrt{\mathrm{Var}(X \cdot 1_A)}\sqrt{\mathrm{Var}(\psi(g))}},$$

it is enough to prove that

$$\mathrm{Corr}\left(X \cdot 1_A, \psi(g)\right) = o(\varepsilon^{(1-\gamma)/2}),$$

since it implies $\mathrm{Cov}(\dots) = o(\varepsilon^{(1-\gamma)/2}) \cdot \|X \cdot 1_A\| = o(\varepsilon^{(1-\gamma)/2}\varepsilon^{\gamma/2}) = o(\sqrt{\varepsilon})$. Instead of $o(\varepsilon^{(1-\gamma)/2})$ we will get $O(\varepsilon^\gamma)$, which is also enough since $\gamma > 1/3$.

It remains to apply the quite general lemma given below, interpreting its Y_k as the whole k-th cell (of g), X_k as the indicator of the event "the k-th cell is not blocked" ($k = 1, \ldots, n$), X_0 as the pair (f, h), and $\varphi(\ldots)$ as $X \cdot 1_A$. The lemma is formulated for real-valued random variables Y_k, but this does not matter; the same clearly holds for arbitrary spaces, and in fact, we need only finite spaces. The product $X_k Y_k$ is a trick for 'blocking' Y_k when $X_k = 0$. Note that dependence between $X_0, X_1, \ldots, X_n$ is allowed.

Lemma 7.4. *Let* $(X_0, X_1, \ldots, X_n)$ *and* $(Y_1, \ldots, Y_n)$ *be two independent random vectors,* $Y_k : \Omega \to \mathbb{R}$, $X_k : \Omega \to \{0, 1\}$ *for* $k = 1, \ldots, n$, $X_0 : \Omega \to \mathbb{R}$, *and random variables* $Y_1, \ldots, Y_n$ *be independent. Then*

$$\mathrm{Corr}\big(\varphi(X_0, X_1 Y_1, \ldots, X_n Y_n),\ \psi(Y_1, \ldots, Y_n)\big) \leq \sqrt{\max_{k=1,\ldots,n} \mathbb{P}\big(X_k = 1\big)}$$

for all Borel functions $\varphi : \mathbb{R}^{n+1} \to \mathbb{R}$, $\psi : \mathbb{R}^n \to \mathbb{R}$ *such that the correlation is well-defined (that is,* $0 < \mathrm{Var}\,\varphi(\ldots) < \infty$, $0 < \mathrm{Var}\,\psi(\ldots) < \infty$*).*

Proof. We may assume that $X_1, \ldots, X_n$ are functions of X_0. Consider the orthogonal (in $L_2(\Omega)$) projection Q from the space of all random variables of the form $\psi(Y_1, \ldots, Y_n)$ to the space of all random variables of the form $\varphi(X_0, X_1 Y_1, \ldots, X_n Y_n)$, that is, $Q\psi(Y_1, \ldots, Y_n) = \mathbb{E}\big(\psi(Y_1, \ldots, Y_n) \,\big|\, X_0, X_1 Y_1, \ldots, X_n Y_n\big)$. We have to prove that $\|Q\psi(Y_1, \ldots, Y_n)\|^2 \leq (\max_k \mathbb{P}(X_k = 1))\|\psi(Y_1, \ldots, Y_n)\|^2$ whenever $\mathbb{E}\,\psi(Y_1, \ldots, Y_n) = 0$. The space of all $\psi(Y_1, \ldots, Y_n)$ is spanned by *factorizable* random variables $\psi(Y_1, \ldots, Y_n) = \psi_1(Y_1) \ldots \psi_n(Y_n)$. For such a ψ we have

$$Q\psi(Y_1, \ldots, Y_n) = \mathbb{E}\big(\psi_1(Y_1) \ldots \psi_n(Y_n) \,\big|\, X_0, X_1 Y_1, \ldots, X_n Y_n\big)$$

$$= \Big(\prod_{k : X_k = 0} \mathbb{E}\,\psi_k(Y_k)\Big)\Big(\prod_{k : X_k = 1} \psi_k(Y_k)\Big);$$

$$\|Q\psi(Y_1, \ldots, Y_n)\|^2 = \mathbb{E}\Big(\mathbb{E}\big(|Q\psi(Y_1, \ldots, Y_n)|^2 \,\big|\, X_0\big)\Big)$$

$$= \mathbb{E}\left(\Big(\prod_{k : X_k = 0} |\mathbb{E}\,\psi_k(Y_k)|^2\Big)\Big(\prod_{k : X_k = 1} \mathbb{E}\,|\psi_k(Y_k)|^2\Big)\right).$$

If, in addition, $\mathbb{E}\,\psi_1(Y_1) = 0$ then $\|Q\psi(Y_1, \ldots, Y_n)\|^2 \leq \mathbb{P}(X_1 = 1)\|\psi(Y_1, \ldots, Y_n)\|^2$. Similarly,

$$\|Q\psi(Y_1, \ldots, Y_n)\|^2 \leq \Big(\max_k \mathbb{P}(X_k = 1)\Big)\|\psi(Y_1, \ldots, Y_n)\|^2$$

if $\mathbb{E}\,\psi(Y_1, \ldots, Y_n) = 0$ and, of course, ψ is factorizable, that is, $\psi(Y_1, \ldots, Y_n) = \psi_1(Y_1) \ldots \psi_n(Y_n)$. The latter assumption cannot be eliminated just by saying that factorizable random variables of zero mean span all random variables of zero mean. Instead, we use two facts.

The first fact. The space of all random variables $\psi(\dots)$ has an *orthogonal* basis consisting of factorizable random variables satisfying an additional condition: each factor $\psi_k(Y_k)$ is either of zero mean, or equal to 1. (For a proof, start with an orthogonal basis for functions of Y_1 only, the first basis function being constant; do the same for Y_2; take all products; and so on.)

The second fact. The operator Q maps *orthogonal* factorizable random variables, satisfying the additional condition, into *orthogonal* random variables. Indeed, let $\psi(Y_1,\dots,Y_n) = \psi_1(Y_1)\dots\psi_n(Y_n)$, $\psi'(Y_1,\dots,Y_n) = \psi_1'(Y_1)\dots\psi_n'(Y_n)$, and each $\psi_k(Y_k)$ be either of zero mean, or equal to 1; the same for each $\psi_k'(Y_k)$. If $\mathbb{E}\big(\psi(Y_1,\dots,Y_n)\psi'(Y_1,\dots,Y_n)\big) = 0$ then $\mathbb{E}\big(\psi_k(Y_k)\psi_k'(Y_k)\big) = 0$ for at least one k; let it happen for $k = 1$. We have not only $\mathbb{E}\big(\psi_1(Y_1)\psi_1'(Y_1)\big) = 0$ but also $\big(\mathbb{E}\,\psi_1(Y_1)\big)\big(\mathbb{E}\,\psi_1'(Y_1)\big) = 0$, since ψ_1 and ψ_1' cannot both be equal to 1. Therefore

$$\mathbb{E}\big(Q\psi(Y_1,\dots,Y_n)\big)\big(Q\psi'(Y_1,\dots,Y_n)\big) =$$

$$= \mathbb{E}\left(\left(\prod_{k:X_k=0}\big(\mathbb{E}\,\psi_k(Y_k)\big)\big(\mathbb{E}\,\psi_k'(Y_k)\big)\right)\left(\prod_{k:X_k=1}\psi_k(Y_k)\psi_k'(Y_k)\right)\right) = 0\,,$$

since the first term vanishes whenever $X_1 = 0$, and the second term vanishes whenever $X_1 = 1$.   □

Combining all together, we get the conclusion.

Theorem 7.5. *The noise of coalescence is black.*

7.4 Remarks

Another proof of Theorem 7.5 should be possible, by showing that all (zero mean) random variables are sensitive. To this end, we divide the time axis $\mathbb{R}$ into intervals of small length ε, and choose a random subset of intervals such that each interval is chosen with a small probability $1 - \rho = 1 - e^{-\lambda} \sim \lambda$, independently of others. On each chosen interval we replace local random data with fresh (independent) data.

Consider the path $X(\cdot)$ of the Brownian web, starting at the origin, $X(t) = \xi_{0,t}(0)$ for $t \in [0,\infty)$; it behaves like a Brownian motion. After the replacement we get another path $Y(\cdot)$. Their difference, $\big(X(t) - Y(t)\big)/\sqrt{2}$, behaves like another Brownian motion when outside 0, but is somewhat sticky at 0. Namely, during each chosen (to the random set) time interval, the point 0 has nothing special; however, outside these time intervals, the point 0 is absorbing. In this sense, chosen time intervals act like factors f_* in the random product of factors f_-, f_+, f_* studied in Sect. 4. There, f_* occurs with a small probability $1/(2\sqrt{i}) \to 0$ (recall Example 4.6), which produces a non-degenerate stickiness in the scaling limit. Here, in contrast, a time interval is chosen with probability $1 - \rho \sim \lambda$ that does not tend to 0 when the interval length ε tends to 0. Naturally, stickiness disappears in the limit $\varepsilon \to 0$ (a proof uses the idea of (4.12)). That is, interaction between $X(\cdot)$ and $Y(\cdot)$ disappears in the limit $\varepsilon \to 0$. They become independent, no matter how small $1 - \rho$ is.

Probably, the same argument works for any finite number of paths $X_k(t) = \xi_{0,t}(x_k)$; they should be asymptotically independent of $Y_k(\cdot)$ for $\varepsilon \to 0$, but I did not prove it.

The spectral measure μ_X of the random variable $X = \xi_{0,1}(0)$ is written down explicitly in [16]. Or rather, its discrete counterpart is found; the scaling limit follows by (a generalization of) Theorem 3.14 (see also [17]). The measure μ_X is a probability measure (since $\|X\| = 1$), it may be thought of as the distribution of a random perfect subset of $(0, 1)$. Note that the random subset is not at all a function on the probability space $(\Omega, \mathcal{F}, P)$ that carries the Brownian web. There is no sense in speaking about 'the joint distribution of the random set and the Brownian web'. In fact, they may be treated as incompatible (non-commuting) measurements in the framework of quantum probability, see [15].

A wonder: μ_X is the distribution of $(\theta - M) \cap (0, 1)$, where M is the set of zeros of the usual Brownian motion, and θ is independent of M and distributed uniformly on $(0, 1)$.

Moreover, the corresponding equality holds exactly (not only asymptotically) in the discrete-time model. Strangely enough, the Brownian motion (or rather, random walk) does not appear in the calculation of the spectral measure. The relation to Brownian motion is observed at the end, as a surprise!

Question 7.6. Can μ_X (for $X = \xi_{0,1}(0)$) be found via some natural construction of a Brownian motion whose zeros form the spectral set (after the transformation $x \mapsto \theta - x$)? (See [16, Problem 1.5].)

We see that μ_X (for $X = \xi_{0,1}(0)$) is concentrated on sets of Hausdorff dimension $1/2$.

Question 7.7. Is μ_X concentrated on sets of Hausdorff dimension $1/2$ for an arbitrary random variable X such that $\mathbb{E} X = 0$ (over the noise of coalescence)?

An affirmative answer would probably give us another proof that the noise is black. A stronger conjecture may be made.

Question 7.8. Is μ_X for an arbitrary $\mathcal{F}_{0,1}$-measurable X (over the noise of coalescence), satisfying $\mathbb{E} X = 0$, absolutely continuous w.r.t. $\mu_{\xi_{0,1}(0)}$?

7.5 A Combinatorial By-product

Consider a Markov chain $X = (X_k)_{k=0}^{\infty}$ (a half-difference of two independent simple random walks, or a double-speed simple random walk divided by two): $X_0 = 0$ and

$$\mathbb{P}\left(X_{k+1} = X_k + \Delta x \mid X_k\right) = \begin{cases} 1/4 & \text{for } \Delta x = -1, \\ 1/2 & \text{for } \Delta x = 0, \\ 1/4 & \text{for } \Delta x = +1 \end{cases}$$

for each $k = 0, 1, 2, \ldots$

Let Z be the (random) set of zeros of X, that is,

$$Z = \{k = 0, 1, \ldots : X_k = 0\}.$$

Given a set $S \subset \{0, 1, 2, \ldots\}$ and a number $k = 0, 1, 2, \ldots$, we consider the event $Z \cap [0, k] \subset k - S$, that is, $\forall l = 0, \ldots, k \; (l \in Z \implies k - l \in S)$, and its probability. We define

$$p_{n,S} = \frac{1}{n} \sum_{k=0}^{n-1} \mathbb{P}\left(Z \cap [0, k] \subset k - S\right);$$

of course, only $k \in S$ can contribute (since $0 \in Z$).

On the other hand, we may trap X at 0 on S; that is, given a set $S \subset \{0, 1, 2, \ldots\}$, we introduce another Markov chain $X^{(S)} = (X_k^{(S)})_{k=0}^{\infty}$ such that $X_0^{(S)} = 0$ and for each $k = 0, 1, 2, \ldots$

$$\mathbb{P}\left(X_{k+1}^{(S)} = x + \Delta x \mid X_k^{(S)} = x\right) = \begin{cases} 1/4 & \text{for } \Delta x = -1, \\ 1/2 & \text{for } \Delta x = 0, \\ 1/4 & \text{for } \Delta x = +1 \end{cases}$$

except for the case $k \in S$, $x = 0$,

$$\mathbb{P}\left(X_{k+1}^{(S)} = 0 \mid X_k^{(S)} = 0\right) = 1 \quad \text{if } k \in S.$$

Theorem 7.9. $p_{n,S} = \frac{1}{n} \sum_{k \in S} \mathbb{P}\left(X_k^{(S)} = 0\right)$ *for every* $n = 1, 2, \ldots$ *and* $S \subset \{0, 1, \ldots, n-1\}$.

Example 7.10. Before proving the theorem, consider a special case; namely, let S consist of just a single number s. Then $\mathbb{P}\left(Z \cap [0, k] \subset k - S\right) = \mathbb{P}\left(Z \cap [0, k] \subset \{k - s\}\right)$ vanishes for $k \neq s$. For $k = s$ it becomes $\mathbb{P}\left(Z \cap [0, s] = \{0\}\right) = 2^{-(2s-1)}\left(\binom{2s-2}{s-1} + \binom{2s-2}{s}\right)$. Therefore $p_{n,\{s\}} = \frac{1}{n} 2^{-(2s-1)}\left(\binom{2s-2}{s-1} + \binom{2s-2}{s}\right)$, assuming $s \geq 2$; also, $p_{n,\{0\}} = \frac{1}{n}$ and $p_{n,\{1\}} = \frac{1}{2n}$. On the other hand, $\frac{1}{n} \sum_{k \in S} \mathbb{P}\left(X_k^{(S)} = 0\right) = \frac{1}{n} \mathbb{P}\left(X_s = 0\right) = \frac{1}{n} \cdot 2^{-2s}\binom{2s}{s}$. The equality becomes $\binom{2s-2}{s-1} + \binom{2s-2}{s} = \frac{1}{2}\binom{2s}{s}$ (for $s \geq 2$).

Proof (sketch). We use the discrete-time counterpart of the Brownian web (see Sect. 7.1 and [16, Sect. 1]) and consider $\xi_{0,n}(0)$, the value at time n of the path starting at the origin. At every instant $k \notin S$ we replace the corresponding random signs with fresh (independent) copies, which leads to another random variable $\xi'_{0,n}(0)$. We calculate the covariance $\mathbb{E}\left(\xi_{0,n}(0)\xi'_{0,n}(0)\right)$ in two ways, and compare the results.

The first way. The difference process $\xi_{0,\cdot}(0) - \xi'_{0,\cdot}(0)$ is distributed like the process $2X^{(S)}$ (similarly to Sect. 7.4). Thus

$$4\mathbb{E}\left(X_n^{(S)}\right)^2 = \mathbb{E}\left(\xi_{0,n}(0) - \xi'_{0,n}(0)\right)^2 = 2n - 2\mathbb{E}\left(\xi_{0,n}(0)\xi'_{0,n}(0)\right).$$

On the other hand, $\frac{1}{2} - \mathbb{E}\left(X_{k+1}^{(S)}\right)^2 + \mathbb{E}\left(X_k^{(S)}\right)^2 = \frac{1}{2}\mathbb{P}\left(X_k^{(S)} = 0\right)$ if $k \in S$, otherwise 0. Therefore $n - 2\mathbb{E}\left(X_n^{(S)}\right)^2 = \sum_{k\in S}\mathbb{P}\left(X_k^{(S)} = 0\right)$. So,

$$\mathbb{E}\left(\xi_{0,n}(0)\xi'_{0,n}(0)\right) = \sum_{k\in S}\mathbb{P}\left(X_k^{(S)} = 0\right).$$

The second way. In terms of the spectral measure μ of the random variable $\xi_{0,n}(0)$ we have $\mathbb{E}\left(\xi_{0,n}(0)\xi'_{0,n}(0)\right) = \mu\{M : M \subset S\}$. However, the probability measure $\frac{1}{n}\mu$ is equal to the distribution of $(\theta - Z) \cap [0,\infty)$; here Z is (as before) the set of zeros of X, and θ is a random variable independent of Z and distributed uniformly on $\{0,1,\dots,n-1\}$. (See [16, Prop. 1.3], see also [24].) Therefore $\frac{1}{n}\mu\{M : M \subset S\} = \mathbb{P}\left((\theta-Z)\cap[0,\infty) \subset S\right) = \mathbb{P}\left(Z\cap[0,\theta] \subset \theta - S\right) = p_{n,S}$. So,

$$\mathbb{E}\left(\xi_{0,n}(0)\xi'_{0,n}(0)\right) = np_{n,S}.$$

$\square$

Question 7.11. Is there a simpler proof of Theorem 7.9? Namely, can we avoid the spectral measure and its relation to the set of zeros?

A continuous-time counterpart of Theorem 7.9 is left to the reader.

8 Miscellany

8.1 Beyond the One-Dimensional Time

Scaling limits of models driven by *two-dimensional* arrays of random signs are evidently important. The best examples appear in percolation theory. Also the Brownian web is an example and, after all, it may be treated as an oriented percolation.

In such cases, independent sub-σ-fields should correspond to disjoint regions of $\mathbb{R}^2$, not only of the form $(s, t) \times \mathbb{R}$. In fact, a rudimentary use of these can be found in Sect. 7 (recall 'cells' in Sect. 7.3). In general it is unclear what kind of regions can be used; probably, regions with piecewise smooth boundaries always fit, while arbitrary open sets do not fit unless the two-dimensional noise is classical (recall Sect. 6.3).

In spite of the great and spectacular progress of the percolation theory (see for instance [14] and references therein), 'the noise of percolation' is still a dream.

Question 8.1. For the critical site percolation on the triangular lattice, invent an appropriate coarse σ-field, and check two-dimensional counterparts of the two conditions of Definition 3.4 for an appropriate class of two-dimensional domains. Is it possible?

Remark 8.2. Hopefully, the answer is affirmative, that is, the two-dimensional noise of percolation will be defined. Then it should appear to be a (two-dimensional) black noise, due to (appropriately adapted) Corollary 6.32, Lemma 7.1 and (most important) the critical exponent for a small cell of size $\varepsilon \times \varepsilon$ being pivotal [14, Sect. 5, Item 2]. The probability is $O(\varepsilon^{5/4})$, therefore $o(\varepsilon)$. The sum for $\mathbf{H}(f)$ contains $O(1/\varepsilon^2)$ terms, $o(\varepsilon^2)$ each.[33]

Sensitivity of percolation events, disclosed in [2], is micro-sensitivity (recall Sect. 5.3). Existence of the black noise of percolation would mean a stronger property: block sensitivity. (See also [2, Problem 5.4].)

It would be the most important example of a black noise!

For the *general* theory of stability, spectral measures, decomposable processes etc., the dimension of the underlying space is of little importance. Basically, regions must form a Boolean algebra. Such a general approach is used in [20], [18].

Nonclassical factorizations appear already in zero-dimensional 'time', be it a Cantor set, or even a convergent sequence with limit point. For Cantor sets, see [20, Sect. 4]; some interesting models of combinatorial nature, with large symmetry groups (instead of 'time shifts' of a noise) are examined there. For a convergent sequence with limit point, see Chapter 1 here (namely, Example 1.1), and [18, Appendix].

---

[33] Different arguments (especially, Lemma 7.4) are used in Sect. 7, since an infinite two-dimensional spectral set could have a finite one-dimensional projection.

8.2 The 'Wave Noise' Approach

A completely different way of constructing noises is sketched here.

Consider the linear wave equation in dimension $1 + 1$,

$$\left(\frac{\partial^2}{\partial t^2} - \frac{\partial^2}{\partial x^2}\right) u(x,t) = 0\,, \tag{8.1}$$

with initial conditions $u(x,0) = 0$, $u_t(x,0) = f(x)$. Its solution is well-known:

$$u(x,t) = \frac{1}{2} \int_{x-t}^{x+t} f(y)\,dy = \frac{1}{2}F(x+t) - \frac{1}{2}F(x-t)\,,$$

where F is defined by $F'(x) = f(x)$. The formula holds in a generalized sense for nonsmooth F, which covers the following case: $F(x) = B(x) =$ Brownian motion (combined out of two independent branches, on $[0, +\infty)$ and on $(-\infty, 0]$); $f(x) = B'(x)$ is the white noise. The random field on $(-\infty, \infty) \times [0, \infty)$,

$$u(x,t) = \frac{1}{2}B(x+t) - \frac{1}{2}B(x-t)\,, \qquad B = \text{Brownian motion,}$$

is continuous, stationary in x, scaling invariant (for any c the random field $u(cx, ct)/\sqrt{c}$ has the same distribution as $u(x,t)$), satisfies the wave equation (8.1) and the following independence condition:

$u\big|_L$ and $u\big|_R$ are independent,

where $L = \{(x,t) : x < -t < 0\}$, $R = \{(x,t) : x > t > 0\}$.

$$\tag{8.2}$$

The independence is a manifestation of: (1) the independence inherent to the white noise (its integrals over disjoint segments are independent), and (2) the hyperbolicity of the wave equation (propagation speed does not exceed 1).

A solution with such properties is essentially unique. That is, if $u(x,t)$ is a continuous random field on $(-\infty, \infty) \times (0, \infty)$, stationary in x, satisfying the wave equation (8.1) and the independence condition (8.2), then necessarily $u(x,t) = \mu_0 + \mu_1 t + \sigma\big(B(x+t) - B(x-t)\big)$ for a Brownian motion B. Scaling invariance forces $\mu_0 = \mu_1 = 0$.

It is instructive that a wave equation may be used in a non-traditional way. Traditionally, a solution is determined by its initial values. In contrast, the independence condition (8.2), combined with some more conditions, determines a random solution with no help of initial conditions! Not an individual sample function is determined, of course, but its distribution (a probability measure on the space of solutions of the wave equation).

Somebody with no preexisting idea of white noise or Brownian motion can, in principle, use the above approach. Observing that $u(x,0) = 0$ but $u_t(x,0)$ does not exist (in the classical sense), he may investigate $u(x,t)/t$ for $t \to 0$ as a way toward the white noise.

Question 8.3. Can we construct a nonclassical (especially, black) noise, using a nonlinear hyperbolic equation?

I once tried the nonlinear wave equation

$$\left(\frac{\partial^2}{\partial t^2} - \frac{\partial^2}{\partial x^2}\right) u(x,t) = \varepsilon t^{-(3-\varepsilon)/2} \sin\left(t^{-(1+\varepsilon)/2} u(x,t)\right),\qquad (8.3)$$

ε being a small positive parameter. The equation is scaling-invariant: if $u(x,t)$ is a solution, then $u(cx, ct)/c^{(1+\varepsilon)/2}$ is also a solution. We search for a random field $u(t, x)$, continuous, stationary in x, scaling invariant, satisfying (8.3) and the independence condition (8.2). Its behavior for $t \to 0$ should give us a new noise. Does such a random field exist? Is it unique (in distribution)? If the answers are affirmative, then we get a noise,

$\mathcal{F}_{x,y}$ is the σ-field generated by $\{u(z,t) : x+t < z < y-t\}$,

and maybe it is black. However, I did not succeed with it.

A modified 'waive noise' approach was used successfully in [20, Sect. 5], proving, for the first time, the existence of a black noise. The modification is to keep the auxiliary dimension, but make it discrete rather than continuous:

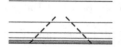

More specifically, consider a sequence of stationary random processes $u_k(\cdot)$ on $\mathbb{R}$ such that

- u_k is $2\varepsilon_k$-dependent (for some $\varepsilon_k \to 0$); it means that $u_k\big|_{(-\infty,-\varepsilon_k]}$ and $u_k\big|_{[\varepsilon_k,+\infty)}$ are independent;

- $u_{k-1}(x)$ is uniquely determined by $u_k\big|_{[x-(\varepsilon_{k-1}-\varepsilon_k),\, x+(\varepsilon_{k-1}-\varepsilon_k)]}$.

Such a sequence (u_k) determines a noise; namely, $\mathcal{F}_{x,y}$ is generated by all $u_k(z)$ such that $x + \varepsilon_k \leq z \leq y - \varepsilon_k$. White noise can be obtained by a linear system of Gaussian processes:

$$u_{k-1}(x) = \int_{x-(\varepsilon_{k-1}-\varepsilon_k)}^{x+(\varepsilon_{k-1}-\varepsilon_k)} V_k(y-x) u_k(y)\, dy,$$

where kernels V_k, concentrated on $[-(\varepsilon_{k-1} - \varepsilon_k), (\varepsilon_{k-1} - \varepsilon_k)]$, are chosen appropriately. A nonlinear system (of quite non-Gaussian processes) of the form

$$u_{k-1}(x) = \varphi\left(\frac{\text{const}}{\varepsilon_{k-1} - \varepsilon_k} \int_{x-(\varepsilon_{k-1}-\varepsilon_k)}^{x+(\varepsilon_{k-1}-\varepsilon_k)} u_k(y)\, dy\right)$$

was used for constructing a black noise. But, it is not really a *construction* of a specific noise. Existence of (u_k) is proven, but uniqueness (in distribution) is not. True, every such (u_k) determines a black noise. However, none of them is singled out.

8.3 Groups, Semigroups, Kernels

A Brownian motion X in a topological group G is defined as a continuous G-valued random process with stationary independent increments, starting from the unit of G. For example, if G is the additive group of reals, then the general form of a Brownian motion in G is $X(t) = \sigma B(t) + vt$, where $B(\cdot)$ is the standard Brownian motion, $\sigma \in [0, \infty)$ and $v \in \mathbb{R}$ are parameters. If G is a Lie group, then Brownian motions X in G correspond to Brownian motions Y in the tangent space of G (at the unit) via the stochastic differential equation $(dX) \cdot X^{-1} = dY$ (in the sense of Stratonovich).

A noise corresponds to every Brownian motion in a topological group, just as the white noise corresponds to $B(\cdot)$. If the noise is classical, it is the white noise of some dimension $(0, 1, 2, \ldots$ or $\infty)$. If this is the case for all Brownian motions in G, we call G a *white group*. Thus, $\mathbb{R}$ is white, and every Lie group is white. Every commutative topological group is white (see [15, Th. 1.8]). The group of all unitary operators in l_2 (equipped with the strong operator topology) is white (see [15, Th. 1.6]). Many other groups are white since they are embeddable into a group known to be white; for example, the group of diffeomorphisms is white (an old result of Baxendale).

Question 8.4. Is the group of all homeomorphisms of (say) $[0, 1]$ white?

In a topological group, Brownian motions X and continuous abstract stochastic flows ξ are basically the same:

$$X(t) = \xi_{0,t}; \qquad \xi_{s,t} = X^{-1}(s)X(t).$$

In a semigroup, however, a noise corresponds to a flow, not to a Brownian motion (see also Example 4.2).

A nonclassical noise (of stickiness) was constructed in Sect. 4 out of an abstract flow in a 3-dimensional semigroup G_3; however, G_3 is not a topological semigroup, since composition is discontinuous.

Question 8.5. Can a nonclassical noise arise from an abstract stochastic flow in a finite-dimensional topological semigroup?

The continuous (but not topological) semigroup G_3 emerged in Sect. 4 from the discrete semigroup G_3^{discrete} via the scaling limit. Or rather, a flow in G_3 emerged from a flow in G_3^{discrete} via the scaling limit. A similar approach to the discrete model of Example 1.9 gives something unexpected. The continuous semigroup that emerges is G_2, the two-dimensional topological semigroup described in (4.15). However, its representation is not single-valued:

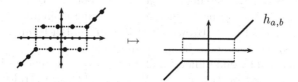

Namely, $h_{a,b}(x)$ for $x \in (-b, b)$ is $\pm(a+b)$, that is, either $a+b$ or $-(a+b)$ with probabilities $0.5, 0.5$. Such h is not a function, of course. Rather, it is a *kernel*, that is, a measurable map from $\mathbb{R}$ into the space of probability measures on $\mathbb{R}$. Composition of kernels is well-defined, thus, a representation (of a semigroup) by kernels (rather than functions) is also well-defined.

The stochastic flow in G_2, resulting from Example 1.9 via the scaling limit, is identical to the flow $(\xi_{s,t}^{(2)})$ of Sect. 4.7. Its noise is the usual (one-dimensional) white noise. The representation of G_2 by kernels turns the abstract flow into a *stochastic flow of kernels* as defined by Le Jan and Raimond [8, Def. 1.1.3]. However, a kernel (unlike a function) introduces an additional level of randomness. When the kernel says that $h_{a,b}(x) = \pm(a + b)$, someone has to choose at random one of the two possibilities. Who makes the decision?

One may treat a point as a macroscopically small collection of many microscopic atoms, and $\omega \in \Omega$ as a macroscopic flow (on the whole space-time); given ω, atoms are (conditionally) independent, "which means that two points[34] thrown initially at the same place separate" [8, p. 4]. No need to deal explicitly with a continuum of independent choices. "Turbulent evolutions [are represented] by flows of probability kernels obtained by dividing infinitely the initial point" [8, p. 4].

Alternatively, one can postulate that if two atoms meet at a (macroscopic!) point, they must coalesce. In one-dimensional space (and *sometimes* in higher dimensions) such a postulate itself prevents a continuum of independent choices and leads to a flow of maps (the Brownian web is an example). A countable dense set of atoms makes decisions; others must obey. A flow of maps is a (degenerate) special case of a flow of kernels. However, coalescence can produce a flow of maps out of a non-degenerate flow of kernels, as explained in [8, Sect. 2.3].

Conversely, a coalescent flow can produce a non-degenerate flow of kernels via "filtering by a sub-noise" [8, Sect. 2.3]. In the simplest case (filtering by a trivial sub-noise), we just retain the one-particle motion of the given coalescent flow, forget the rest of the flow, and let atoms perform the motion independently.

A large class of flows on $\mathbb{R}^n$ (and other homogeneous spaces) is investigated in [8]. Some of these flows are shown to be coalescent and to generate nonclassical noises (neither white nor black). Flows are homogeneous in space (and isotropic). Thus, we have a hierarchy of nonclassical models. First, toy models

---

[34] Or rather, atoms.

(recall Examples 1.1, 1.2) having a singular time point. Second, 'simple' models (Sects. 1.4, 4.9) homogeneous in time but having a singular spatial point. Third, 'serious' models (the Brownian web, and Le Jan-Raimond's isotropic Brownian flows), homogeneous in space and time.

Noises generated by one-dimensional flows (also homogeneous in space and time) are investigated by Warren and Watanabe [24]. Spectral sets of Hausdorff dimension other than 0 and 1/2 are found! Roughly, it answers Question 6.26; however, these spectral sets are not perfect — they have isolated points.

8.4 Abstract Nonsense of Le Jan-Raimond's Theory

A new semigroup, introduced recently by Le Jan and Raimond [8], is quite interesting for the theory of stochastic flows and noises. Its definition involves some technicalities considered here.

A kernel is defined in [8] as a measurable mapping from a compact metric space $\mathcal{M}$ to the (also compact) space $\mathcal{P}(\mathcal{M})$ of all probability measures on $\mathcal{M}$. The space E of all kernels is equipped with the σ-field $\mathcal{E}$ generated by evaluations, $E \ni K \mapsto K(x) \in \mathcal{P}(\mathcal{M})$, at points $x \in \mathcal{M}$. Note that every $\mathcal{E}$-measurable function uses the values of $K(x)$ only for a countable set of points x, which is scanty, since $K(x)$ is just measurable (rather than continuous) in x. Thus, $(E, \mathcal{E})$ is not a standard Borel space,[35] and the composition of kernels is not a measurable operation, which obscures the technique and makes proofs more difficult (as noted on page 11 of [8]).

Fortunately, the theory can be reformulated equivalently in terms of Borel operations on standard Borel spaces, as outlined below. Additional simplification comes from disentangling space and time (entangled in Theorem 1.1.4 of [8]) and explicit use of the de Finetti theorem.

The hassle about measurability is another manifestation of the well-known clash between finite-dimensional distributions and modifications of a random process. Say, for the usual Poisson process on $[0, \infty)$, its finite-dimensional distributions do not tell us whether sample paths are continuous from the left (right), or not. A process $X = X(t, \omega)$ has a lot of modifications $Y(t, \omega)$; these satisfy $\forall t \; \mathbb{P}\left(\{\omega : X(t, \omega) = Y(t, \omega)\}\right) = 1$, which does not imply $\mathbb{P}\left(\{\omega : \forall t \; X(t, \omega) = Y(t, \omega)\}\right) = 1$. If a process admits continuous sample paths (like the Brownian motion), the continuous modification is preferable. If a process is just continuous in probability (like the Poisson process, but also, say, some stationary Gaussian processes, unbounded on every interval), we are unable to prefer one modification to others, in general.

In order to describe the class of all modifications of a random process, we have two well-known tools: first, a compatible family of finite-dimensional distributions, and second, a probability measure on the (non-standard!) Borel space of all (or only measurable; but definitely, not only continuous) sample paths, whose σ-field is generated by evaluations. Assuming the process to be

---

[35] For a definition, see [7, Sect. 12.B] or [1, Def. 7.1].

continuous in probability, we find the first tool much better; joint distributions depend on points continuously, and everything is standard.

The same for kernels. These may be thought of as sample paths of a random process whose 'time' runs over $\mathcal{M}$, and 'values' belong to $\mathcal{P}(\mathcal{M})$. However, the process will appear (implicitly) only in Theorem 8.8; its finite-dimensional distributions are $\nu_n(x_1, \ldots, x_n)$ there.

Definition 8.6. *A* multikernel *from a compact metric space $\mathcal{M}_1$ to a compact metric space $\mathcal{M}_2$ is a sequence $(P_n)_{n=1}^{\infty}$ of continuous maps $P_n : \mathcal{M}_1^n \to \mathcal{P}(\mathcal{M}_2^n)$, compatible in the sense that[36]*

$$\int_{\mathcal{M}_2^n} g \, dP_n(x_1, \ldots, x_n) = \int_{\mathcal{M}_2^m} f \, dP_m(x_{i_1}, \ldots, x_{i_m})$$

for all n and $x_1, \ldots, x_n \in \mathcal{M}_1$, whenever $i_1, \ldots i_m$ are pairwise distinct elements of $\{1, \ldots, n\}$, $f : \mathcal{M}_2^m \to \mathbb{R}$ is a continuous function, and $g : \mathcal{M}_2^n \to \mathbb{R}$ is defined by $g(y_1, \ldots, y_n) = f(y_{i_1}, \ldots, y_{i_m})$ for $y_1, \ldots, y_n \in \mathcal{M}_2$.

We do not assume $i_1 < \cdots < i_m$. For example:

$$g(y_1, y_2) = f(y_1) \quad \Longrightarrow \quad \int g \, dP_2(x_1, x_2) = \int f \, dP_1(x_1);$$

$$g(y_1, y_2) = f(y_2) \quad \Longrightarrow \quad \int g \, dP_2(x_1, x_2) = \int f \, dP_1(x_2);$$

$$g(y_1, y_2) = f(y_2, y_1) \quad \Longrightarrow \quad \int g \, dP_2(x_1, x_2) = \int f \, dP_2(x_2, x_1).$$

Note also that $x_1, x_2, \ldots$ need not be distinct.

Definition 8.7. *A* multikernel *$(P_n)_{n=1}^{\infty}$ is* single-valued, *if*

$$\int_{\mathcal{M}_2^2} g \, dP_2(x, x) = \int_{\mathcal{M}_2} f \, dP_1(x) \quad \text{for all } x \in \mathcal{M}_1,$$

whenever $g : \mathcal{M}_2^2 \to \mathbb{R}$ is a continuous function, and $f : \mathcal{M}_2 \to \mathbb{R}$ is defined by $f(y) = g(y, y)$ for $y \in \mathcal{M}_2$.

An equivalent definition: $(P_n)_{n=1}^{\infty}$ is single-valued, if

$$\int_{\mathcal{M}_2^2} \rho \, dP_2(x, x) = 0 \quad \text{for all } x \in \mathcal{M}_1,$$

where $\rho : \mathcal{M}_2^2 \to \mathbb{R}$ is the metric, $\rho(y_1, y_2) = \mathrm{dist}(y_1, y_2)$.

Another equivalent definition:

---

[36] Here $\int g \, dP_n(x_1, \ldots, x_n)$ is not an integral in $x_1, \ldots, x_n$. Rather, $x_1, \ldots, x_n$ are parameters. The integral is taken in other variables (say, $y_1, \ldots, y_n$), suppressed in the notation and running over $\mathcal{M}_2$.

$$\sup_{\rho(x_1,x_2)\le\varepsilon}\int_{\mathcal{M}_2^2}\rho\,\mathrm{d}P_2(x_1,x_2)\to 0\quad\text{for }\varepsilon\to 0\,.$$

(Compare it with continuity in probability.)

My 'multikernel' is a time-free counterpart of a 'compatible family of Feller semigroups' of [8]. My 'single-valued' corresponds to their (1.7). What could correspond to their 'stochastic convolution semigroup'? It is a single-valued multikernel from $\mathcal{M}_1$ to $\mathcal{P}(\mathcal{M}_2)$. Yes, I mean it: maps from $\mathcal{M}_1^n$ to $\mathcal{P}\big((\mathcal{P}(\mathcal{M}_2))^n\big)$. It may look frightening, but think what happens if $\mathcal{M}_1$ contains only one point, and $\mathcal{M}_2$ — only two points, say, 0 and 1. Then a multikernel from $\mathcal{M}_1$ to $\mathcal{M}_2$ is a law of an exchangeable sequence of events. A single-valued multikernel from $\mathcal{M}_1$ to $\mathcal{M}_2$ would mean that all events coincide, but we need rather a single-valued multikernel from $\mathcal{M}_1$ to $\mathcal{P}(\mathcal{M}_2)=[0,1]$; nothing but a probability measure on $[0,1]$. The De Finetti theorem (see [1], for instance) tells us that every exchangeable sequence of events arises from a probability measure on $[0,1]$. Here is a more general result.

Theorem 8.8. *For every multikernel $(P_n)_{n=1}^\infty$ from $\mathcal{M}_1$ to $\mathcal{M}_2$ there exists a single-valued multikernel $(\nu_n)_{n=1}^\infty$ from $\mathcal{M}_1$ to $\mathcal{P}(\mathcal{M}_2)$ such that*

$$\int_{\mathcal{M}_2^n} f\,\mathrm{d}P_n(x_1,\dots,x_n)=\int_{(\mathcal{P}(\mathcal{M}_2))^n}F\,\mathrm{d}\nu_n(x_1,\dots,x_n)$$

for all n and $x_1,\dots,x_n\in\mathcal{M}_1$, whenever $f:\mathcal{M}_2^n\to\mathbb{R}$ is a continuous function, and $F:(\mathcal{P}(\mathcal{M}_2))^n\to\mathbb{R}$ is defined by $F(\mu_1,\dots,\mu_n)=\int f\,d(\mu_1\otimes\cdots\otimes\mu_n)$ for $\mu_1,\dots,\mu_n\in\mathcal{P}(\mathcal{M}_2)$.

Proof. We choose a discrete probability measure μ_0 on $\mathcal{M}_1$ whose support is the whole $\mathcal{M}_1$. That is, we choose a countable (or finite) dense set $A\subset\mathcal{M}_1$, and give a positive probability to each point of A. For every n we consider the following measure Q_n on $(\mathcal{M}_1\times\mathcal{M}_2)^n$:

$$\int f_1\otimes g_1\otimes\cdots\otimes f_n\otimes g_n\,\mathrm{d}Q_n$$

$$=\int\left(\int g_1\otimes\cdots\otimes g_n\,\mathrm{d}P_n(x_1,\dots,x_n)\right)f_1(x_1)\dots f_n(x_n)\,\mathrm{d}\mu_0(x_1)\dots\mathrm{d}\mu_0(x_n)\,.$$

In other words, if Q_n is the distribution of $(X_1,Y_1;\dots;X_n,Y_n)$, then $X_1,\dots,X_n$ are i.i.d. distributed μ_0 each, and the conditional distribution of $(Y_1,\dots,Y_n)$ given $(X_1,\dots,X_n)$ is $P_n(X_1,\dots,X_n)$. The measure Q_n is invariant under the group of $n!$ permutations of n pairs, due to compatibility of the multikernel $(P_n)_{n=1}^\infty$. For the same reason, Q_n is the marginal of Q_{n+1}. Thus, $(Q_n)_{n=1}^\infty$ is the distribution of an exchangeable infinite sequence of $\mathcal{M}_1\times\mathcal{M}_2$-valued random variables (X_n,Y_n).

The De Finetti theorem [1, Th. 3.1 and Prop. 7.4] states that the joint distribution of all (X_n,Y_n) is a mixture of products, in the sense that there exists a probability measure ν on $\mathcal{P}(\mathcal{M}_1 \times \mathcal{M}_2)$ such that for every n, the joint distribution of n pairs $(X_1, Y_1), \ldots, (X_n, Y_n)$ is the mixture of products $Q^{\otimes n} = Q \otimes \cdots \otimes Q$, where $Q \in \mathcal{P}(\mathcal{M}_1 \times \mathcal{M}_2)$ is distributed ν. The first marginal of Q is equal to μ_0 (for ν-almost every Q), since X_n are i.i.d. (μ_0).

Let $x_1, \ldots, x_n \in A$. The event $X_1 = x_1, \ldots, X_n = x_n$ is of positive probability. Given the event, the conditional distribution $P_n(x_1, \ldots, x_n)$ of $Y_1, \ldots, Y_n$ is the mixture of products $Q_{x_1} \otimes \cdots \otimes Q_{x_n}$, where Q_x is the conditional measure on $\mathcal{M}_2$, that corresponds to Q, and $Q \in \mathcal{P}(\mathcal{M}_1 \times \mathcal{M}_2)$ is distributed ν; indeed, ν-almost all Q ascribe the same probability to the event $X_1 = x_1, \ldots, X_n = x_n$.

We define $\nu_n(x_1, \ldots, x_n)$ for $x_1, \ldots, x_n \in A$ as the joint distribution of $\mathcal{P}(\mathcal{M}_2)$-valued random variables $Q_{x_1}, \ldots, Q_{x_n}$, where Q is distributed ν; then

$$\int_{(\mathcal{P}(\mathcal{M}_2))^n} F \, d\nu_n(x_1, \ldots, x_n)$$

$$= \int_{\mathcal{P}(\mathcal{M}_1 \times \mathcal{M}_2)} \left(\int_{\mathcal{M}_2^n} f \, d(Q_{x_1} \otimes \cdots \otimes Q_{x_n}) \right) d\nu(Q)$$

$$= \int_{\mathcal{M}_2^n} f \, dP_n(x_1, \ldots, x_n) \quad (8.4)$$

whenever $f : \mathcal{M}_2^n \to \mathbb{R}$ is a continuous function, and $F : (\mathcal{P}(\mathcal{M}_2))^n \to \mathbb{R}$ is defined by $F(\mu_1, \ldots, \mu_n) = \int f \, d(\mu_1 \otimes \cdots \otimes \mu_n)$ for $\mu_1, \ldots, \mu_n \in \mathcal{P}(\mathcal{M}_2)$.

Till now, $\nu_n(x_1, \ldots, x_n)$ is defined for $x_1, \ldots, x_n \in A$ (rather than $\mathcal{M}_1$). We want to check that $\int \tilde{\rho}_2 \, d\nu_2(x_1, x_2) \to 0$ for $\rho_1(x_1, x_2) \to 0$; here ρ_1 is a metric on $\mathcal{M}_1$ conforming to its topology, and $\tilde{\rho}_2$ is a metric on $\mathcal{P}(\mathcal{M}_2)$ conforming to its weak topology. Due to compactness of $\mathcal{P}(\mathcal{M}_2)$, it is enough to check that $\int h^2 \, d\nu_2(x_1, x_2) \to 0$ for $\rho_1(x_1, x_2) \to 0$ whenever $h : \mathcal{P}(\mathcal{M}_2) \times \mathcal{P}(\mathcal{M}_2) \to \mathbb{R}$ is of the form $h(Q_1, Q_2) = \int f \, dQ_1 - \int f \, dQ_2$ for a continuous function $f : \mathcal{M}_2 \to \mathbb{R}$. Consider $\tilde{f} : \mathcal{P}(\mathcal{M}_2) \to \mathbb{R}$, $\tilde{f}(Q) = \int f \, dQ$ for $Q \in \mathcal{P}(\mathcal{M}_2)$. We have

$$\int_{(\mathcal{P}(\mathcal{M}_2))^2} \tilde{f} \otimes \tilde{f} \, d\nu_2(x_1, x_2) = \int_{\mathcal{M}_2^2} f \otimes f \, dP_2(x_1, x_2),$$

which is a special case of (8.4). It may also be written as

$$\mathbb{E} \, \tilde{f}(Q_{x_1}) \tilde{f}(Q_{x_2}) = \mathbb{E} \left(f(Y_1) f(Y_2) \,|\, X_1 = x_1, X_2 = x_2 \right);$$

here Q_{x_1} and Q_{x_2} are treated as random variables on the probability space $\left(\mathcal{P}(\mathcal{M}_1 \times \mathcal{M}_2), \nu \right)$ (thus, the two expectations are taken on different probability spaces). The right-hand side is a continuous function of x_1, x_2; denote it $\varphi(x_1, x_2)$. We have

$$\int h^2 \, d\nu_2(x_1, x_2) = \mathbb{E} \left(\tilde{f}(Q_{x_1}) - \tilde{f}(Q_{x_2}) \right)^2$$

$$= \varphi(x_1, x_1) - \varphi(x_1, x_2) - \varphi(x_2, x_1) + \varphi(x_2, x_2),$$

which tends to 0 for $\rho_1(x_1, x_2) \to 0$. So,

$$\int_{(\mathcal{P}(\mathcal{M}_2))^2} \tilde{\rho}_2 \, d\nu_2(x_1, x_2) \to 0 \quad \text{for } \rho_1(x_1, x_2) \to 0 \, .$$

It follows easily that each ν_n is uniformly continuous on A^n and, extending it by continuity to $\mathcal{M}_1^n$, we get a single-valued multikernel.   □

Definition 8.6 may be reformulated as follows.

Definition 8.9. *A multikernel from a compact metric space $\mathcal{M}_1$ to a compact metric space $\mathcal{M}_2$ is a continuous map $P_\infty : \mathcal{M}_1^\infty \to \mathcal{P}(\mathcal{M}_2^\infty)$, satisfying conditions (1) and (2) below. Here $\mathcal{M}^\infty = \mathcal{M} \times \mathcal{M} \times \ldots$ is the product of an infinite sequence of copies of $\mathcal{M}$ (still a metrizable compact space).*

(1) P_∞ intertwines the natural actions of the permutation group of the index set $\{1, 2, 3, \ldots\}$ on $\mathcal{M}_1^\infty$ and $\mathcal{P}(\mathcal{M}_2^\infty)$ (via $\mathcal{M}_2^\infty$).

(2) For every n, the projection of the measure $P_\infty(m)$ to the product $\mathcal{M}_1^n$ of the first n factors depends only on the first n coordinates $m_1, \ldots, m_n$ of the point $(m_1, m_2, \ldots) = m \in \mathcal{M}_1^\infty$.

Proof of equivalence between definitions 8.6 and 8.9 is left to the reader.

It is well-known that a continuous map $\mathcal{M}_1 \to \mathcal{P}(\mathcal{M}_2)$ is basically the same as a linear operator $C(\mathcal{M}_2) \to C(\mathcal{M}_1)$, positive and preserving the unit. Thus, a multikernel from $\mathcal{M}_1$ to $\mathcal{M}_2$ may be thought of as a positive unit-preserving linear operator $C(\mathcal{M}_2^\infty) \to C(\mathcal{M}_1^\infty)$ satisfying two conditions parallel to 8.9(1,2).

Given three compact metric spaces $\mathcal{M}_1, \mathcal{M}_2, \mathcal{M}_3$, a multikernel from $\mathcal{M}_1$ to $\mathcal{M}_2$ and a multikernel from $\mathcal{M}_2$ to $\mathcal{M}_3$, we may define their composition, a multikernel from $\mathcal{M}_1$ to $\mathcal{M}_3$. In terms of operators it is just the product of two operators, $C(\mathcal{M}_3^\infty) \to C(\mathcal{M}_2^\infty) \to C(\mathcal{M}_1^\infty)$.

The set of all multikernels from $\mathcal{M}_1$ to $\mathcal{M}_2$, treated as operators $C(\mathcal{M}_2^\infty) \to C(\mathcal{M}_1^\infty)$, is a closed (and bounded, but not compact) subset of the operator space equipped with the strong operator topology. Thus, the set of multikernels becomes a Polish space (that is, a topological space underlying a complete separable metric space).

Composition of multikernels, $C(\mathcal{M}_3^\infty) \to C(\mathcal{M}_2^\infty) \to C(\mathcal{M}_1^\infty)$, is a (jointly) continuous operation. (Indeed, the product of operators is continuous in the strong operator topology, as far as all operators are of norm ≤ 1.)

So, multikernels from $\mathcal{M}$ to $\mathcal{M}$ are a Polish semigroup (that is, a topological semigroup whose topological space is Polish).

References

1. Aldous, D.J. (1985): Exchangeability and related topics. In: Lecture Notes in Math. **1117** (École de Saint-Flour XIII), 1–198.
2. Benjamini, I., Kalai, G., Schramm, O. (1999): Noise sensitivity of Boolean functions and applications to percolation. Inst. Hautes Études Sci. Publ. Math. no. 90, 5–43.
3. Émery, M., Schachermayer, W. (1999): A remark on Tsirelson's stochastic differential equation. In: Lecture Notes in Math. **1709** (Séminaire de Probabilités XXXIII), 291–303.
4. Feldman, J. (1971): Decomposable processes and continuous products of probability spaces. J. Funct. Anal. **8**, 1–51.
5. Fontes, L.R.G., Isopi, M., Newman, C.M., Ravishankar, K. (2002): The Brownian web. arXiv:math.PR/0203184.
6. Hawkes, J. (1981): Trees generated by a simple branching process. J. London Math. Soc. (2) **24**, 373–384.
7. Kechris, A.S. (1995): Classical Descriptive Set Theory. Springer Berlin Heidelberg.
8. Le Jan, Y., Raimond, O. (2002): Flows, coalescence and noise. arXiv:math.PR/0203221.
9. Le Jan, Y., Raimond, O. (2002): The noise of a Brownian sticky flow is black. arXiv:math.PR/0212269 (v1).
10. Peres, Y. (1996): Intersection equivalence of Brownian paths and certain branching processes. Commun. Math. Phys. **177**, 417–434.
11. Revuz, D., Yor, M. (1994): Continuous Martingales and Brownian Motion. Second edition. Springer Berlin Heidelberg.
12. Schramm, O., Tsirelson, B. (1999): Trees, not cubes: hypercontractivity, cosiness, and noise stability. Electronic Communications in Probability, **4**, 39–49.
13. Shnirelman, A. (1997): On the nonuniqueness of weak solution of the Euler equation. Comm. Pure Appl. Math., **50**:12, 1261–1286.
14. Smirnov, S., Werner, W. (2001): Critical exponents for two-dimensional percolation. Mathematical Research Letters, **8**, 729–744.
15. Tsirelson, B. (1998): Unitary Brownian motions are linearizable. arXiv:math.PR/9806112.
16. Tsirelson, B. (1999): Fourier-Walsh coefficients for a coalescing flow (discrete time). arXiv:math.PR/9903068.
17. Tsirelson, B. (1999): Scaling limit of Fourier-Walsh coefficients (a framework). arXiv:math.PR/9903121.
18. Tsirelson, B. (1999): Noise sensitivity on continuous products: an answer to an old question of J. Feldman. arXiv:math.PR/9907011.
19. Tsirelson, B. (2002): Non-isomorphic product systems. arXiv:math.FA/0210457. To be publ. in: Advances in Quantum Dynamics (eds. G. Price et al), "Contemporary Mathematics", AMS.
20. Tsirelson, B.S., Vershik, A.M. (1998): Examples of nonlinear continuous tensor products of measure spaces and non-Fock factorizations. Reviews in Mathematical Physics **10**:1, 81–145.
21. Warren, J. (1997): Branching processes, the Ray-Knight theorem, and sticky Brownian motion. In: Lecture Notes in Math. **1655** (Séminaire de Probabilités XXXI), 1–15.

22. Warren, J. (1999): Splitting: Tanaka's SDE revisited. arXiv:math.PR/9911115.
23. Warren, J. (2002): The noise made by a Poisson snake. Electronic Journal of Probability **7**:21, 1–21.
24. Warren, J., Watanabe, S.: On Harris's stochastic flows. (In preparation.)
25. Watanabe, S. (2000): The stochastic flow and the noise associated to Tanaka's stochastic differential equation. Ukrainian Math. J. **52**:9, 1346–1365 (transl).
26. Watanabe, S. (2001): A simple example of black noise. Bull. Sci. Math. **125**:6/7, 605–622.
27. v. Weizsäcker, H. (1983): Exchanging the order of taking suprema and countable intersections of sigma-algebras. Ann. Inst. Henri Poincaré B **19**:1, 91–100.

Index

Part II

Wendelin Werner: Random Planar Curves and
Schramm-Loewner Evolutions

Random Planar Curves and Schramm-Loewner Evolutions

Wendelin Werner

Université Paris-Sud and IUF
Laboratoire de Mathématiques, Université Paris-Sud,
Bât. 425, 91405 Orsay cedex, France
e-mail: wendelin.werner@math.u-psud.fr

Foreword and Summary

The goal of these lectures is to review some of the mathematical results that have been derived in the last years on conformal invariance, scaling limits and properties of some two-dimensional random curves. The (distinguished) audience of the Saint-Flour summer school consists mainly of probabilists and I therefore assume knowledge in stochastic calculus (Itô's formula etc.), but no special background in basic complex analysis.

These lecture notes are neither a book nor a compilation of research papers. While preparing them, I realized that it was hopeless to present all the recent results on this subject, or even to give the complete detailed proofs of a selected portion of them. Maybe this will disappoint part of the audience but the main goal of these lectures will be to try to transmit some ideas and heuristics. As a reader/part of an audience, I often think that omitting details is dangerous, and that ideas are sometimes better understood when the complete proofs are given, but in the present case, partly because the technicalities often use complex analysis tools that the audience might not be so familiar with, partly also because of the limited number of lectures, I chose to focus on some selected results and on the main ideas of their proofs, sometimes omitting technical details, and giving references for those interested in full proofs or more results. In the final chapter, I will briefly review what I omitted in these lectures, as well as work in progress or open questions.

Of course, I would like to thank my coauthors Greg Lawler and Oded Schramm without which I would not have been lecturing on this subject in Saint-Flour. Collaborating with them during these last years was a great pleasure and privilege. Also, I would like to stress the fact that (almost) none of the pictures in these notes are mine. Many thanks to their authors Vincent Beffara, Tom Kennedy and Oded Schramm. I also take this opportunity to thank Stas Smirnov, Rick Kenyon, as well as all my Orsay colleagues and students who have directly or indirectly contributed to these lecture notes through their work, comments and discussions.

Finally, I owe many thanks to all participants of the summer school, as well as to all colleagues who have sent me their comments and remarks on the first draft of these notes that was distributed during the summer school and posted on the web at that time.

It has been a pleasure and a very rewarding experience to lecture in the studious, relaxed and enjoyable atmosphere of the 2002 St-Flour school. I express my gratitude to all who have contributed to it, my co-lecturers Jim Pitman and Boris Tsirelson, the Maison des Planchettes' staff, and last but not least, Jean Picard, whose outstanding organization has been both efficient and discreet.

Here is a short description of these notes: In the first introductory chapter, I will briefly describe two discrete models (loop-erased random walks and critical percolation interfaces) that have now been proved to converge in their scaling limit to SLE (Oded Schramm used these letters as shorthand for "stochastic Loewner Evolution", but I will stick to Schramm-Loewner Evolution). Using these models, I will try to show why it is natural to define this one-parameter family of random continuously growing processes based on Loewner's equation, and to introduce the difference between their chordal and radial versions.

The second chapter is a review of the necessary background on deterministic aspects of Loewner's equation in the upper half-plane, which is then used in Chapter 3 to define chordal SLE. Some first properties of this process are studied. In particular, some hitting probabilities are computed.

The fourth chapter is devoted to some special properties of SLE that hold for some special values of the parameter κ: The locality property for SLE_6, and the restriction property for $SLE_{8/3}$. These are not surprising if one thinks of these processes as the respective scaling limits of critical percolation interfaces and self-avoiding walks, but somewhat surprising if one starts from the definition of SLE itself. These properties are then used in Chapter 5, to make the link between the geometry of $SLE_{8/3}$, that of the outer boundary of a planar Brownian motion and that of the outer boundary of SLE_6.

In Chapter 6, we define radial SLE which are processes defined in a similar way as chordal SLE except that they are growing towards an interior point of the domain and not to a boundary point. We show in that chapter that radial and chordal SLE are very closely related, especially in the case $\kappa = 6$.

In Chapter 7, we show how to compute critical exponents associated to SLE that describe the asymptotic decay of certain probabilities (non-disconnection, non-intersection). Using the relation between radial SLE_6, chordal SLE_6 and planar Brownian motion, we then use these computations in Chapter 8 to determine the values of the critical exponents that describe the decay of disconnection or non-intersection probabilities for planar Brownian motions, which is one of the main goals of these lectures. As already mentioned, it will not be possible to describe all proofs in detail, but I hope that all the main ideas and steps (that are spread over the first seven chapters of these notes) are explained in sufficient detail so that the reader can get an overview of the proof. For simplicity, I will mainly focus on derivation of the disconnection exponent i.e. the proof of the fact that the probability that a complex Brownian curve $Z[0, t]$ started from $Z_0 = 1$ disconnects the origin from infinity decays like $t^{-1/8}$ when $t \to \infty$.

In Chapters 9 and 10, another important aspect of SLE is discussed: The proofs that some curves arising in discrete models from statistical physics converge to SLE in their scaling limit. The case of loop-erased random walks and uniform spanning trees is treated in Chapter 9. Chapter 10 is devoted to critical site percolation on the triangular lattice, including a brief discussion of Stas Smirnov's proof of conformal invariance and of its consequences.

A concluding chapter contains a list of other results, work in progress and open questions.

1 Introduction

1.1 General Motivation

One of the main aims of both statistical physics and probability theory is to study macroscopic systems consisting of many (i.e. in the limit when this number grows to infinity) small microscopic random inputs. One may classify the results into two categories: In the limit, the behaviour of the macroscopic system becomes deterministic (these are "law of large number" type of results, and large deviations can to some extent been used in this framework), or random. The archetype for continuous random objects that appear as scaling limit of finite systems is Brownian motion. Note that it is the scaling limit of a large class of simple random walks, so that one might argue that Brownian motion is more universal than the discrete model (simple random walk) because there is no need to specify a lattice or a jump-distribution: it only captures the phenomenological properties of the walks (mean zero, stationary increments etc.).

In two dimensions, Brownian motion has an important property which was first observed by Paul Lévy ([102], see e.g. [100, 117] for "modern" proofs based on Itô's formula) and that can be heuristically related to the fact that it is the scaling limit of simple random walks on different lattices (which implies for instance invariance under rotations and under scaling): It is invariant under conformal transformations. Here is one way to state this property: Take a simply connected open planar domain D that contains the origin and is not equal to $\mathbb{C}$. Consider planar Brownian motion $(B_t, t \in [0, \tau])$ started from $B_0 = 0$ up to its exit time $\tau = \tau_D$ of the domain D. Suppose that Φ is a conformal map (that is, a one-to-one smooth map that preserves angles) from D onto some other domain D' with $\Phi(0) = 0$. Then, there exists a (random) time change $A : [0, \sigma] \to [0, \tau]$ so that $(\Phi(B_{A(s)}), s \in [0, \sigma])$ is planar Brownian motion started from 0 and killed at its first exit time σ of D'. In other words, if we forget about time-parametrization, the law of $\Phi(B)$ is again a Brownian motion. As we shall see in these lectures, conformal invariance will turn out to be instrumental in the understanding of curves arising in more complicated setups.

Actually, there exist only few known examples of probabilistic continuous models that are not directly related to Brownian motion. For instance, under mild regularity conditions, continuous finite-dimensional Markov processes are solutions of stochastic differential equations and therefore constructed using Brownian motions. If one looks for other types of continuous processes, one has therefore to give up the Markov property or the finite-dimensionality. In many complex systems that we see around us and for which probability theory seems a priori a well-suited tool (the shape of clouds, say), it is not possible to explain the phenomena via Brownian motions, and there is still a long way to go for probabilists to understand their macroscopic behaviour.

In the present lectures, we shall focus on random planar curves. In two dimensions, (random) curves appear naturally as boundaries of domains, in-

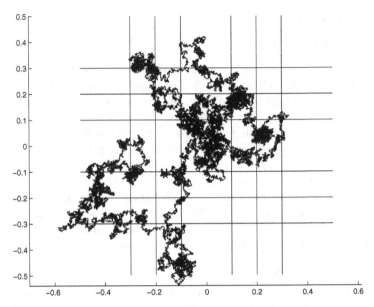

Fig. 1.1. Sample of a long simple random walk.

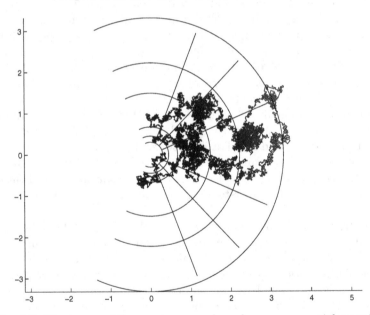

Fig. 1.2. The image of the previous sample under an exponential mapping.

terfaces between two phases, level lines of random surfaces etc. In all these cases, at least on microscopic level, the definition of the curve (say, as an interface) implies that it is a self-avoiding curve (or a simple closed loop). On the macroscopic scale, the continuous curves that we will be considering may have double-points (in the scaling limit, simple curves may converge to curves with multiple points), but self-crossings are forbidden. Of course, if $(\gamma_t, t \in [0, T])$ is such a random curve, we see that in general, this condition implies a strong correlation between $\gamma[0, t]$ and $\gamma[t, T]$, so that the Markov property is lost (if we look at these curves as living in the two-dimensional space). As we shall see, there is a way to recover a Markov property for the random curves, using a coding of the curve in an infinite-dimensional space of conformal maps.

1.2 Loop-Erased Random Walks

In order to guide the intuition about the family of random curves that we will be considering, it is helpful to have some discrete models in mind, for which one expects or can prove that they converge to this continuous object. We therefore start these lectures with the description of one measure on discrete random curves that turns out to converge in the scaling limit. This is actually the model that Oded Schramm considered when he invented these random curves that he called *SLE* (for Stochastic Loewner Evolution, but we will replace this by Schramm-Loewner Evolution in these lectures).

For any $\mathbf{x} = (x_0, \ldots, x_m)$, we define the loop-erasure $L(\mathbf{x})$ of $\mathbf{x}$ inductively as follows: $L_0 = x_0$, and for all $j \geq 0$, we define inductively $n_j = \max\{n \leq m : x_n = L_j\}$ and

$$L_{j+1} = X_{1+n_j}$$

until $j = \sigma$ where $L_\sigma := x_m$. In other words, we have erased the loops of $\mathbf{x}$ in chronological order. The number of steps σ of L is not fixed.

Suppose that $(X_n, n \geq 0)$ is a recurrent Markov chain on a discrete statespace $\mathcal{S}$ started from $X_0 = x$. Suppose that $A \subset \mathcal{S}$ is non-empty, and let τ_A denote the hitting time of A by X. Let $p(x, y)$ denote the transition probabilities for the Markov chain X. We define the loop-erasure $L = L(X[0, \tau_A]) = L^A$ of X up to its hitting time of A. We call σ the number of steps of L^A. For $y \in A$ such that with positive probability $L^A(\sigma) = X(\tau_A) = y$, we call $\mathcal{L}(x, y; A)$ the law of L^A conditioned on the event $\{L^A(\sigma) = y\}$. In other words, it is the law of the loop-erasure of the Markov chain X conditioned to hit A at y.

Lemma 1.1 (Markovian property of LERW). *Consider* $y_0, \ldots, y_j \in \mathcal{S}$ *so that with positive probability for* $\mathcal{L}(x, y_0; A)$,

$$\{L_\sigma = y_0, L_{\sigma-1} = y_1, \ldots, L_{\sigma-j} = y_j\}.$$

The conditional law of $L[0, \sigma - j]$ *given this event is* $\mathcal{L}(x, y_j; A \cup \{y_1, \ldots, y_j\})$.

Proof. For each A and $x \in \mathcal{A}$, we denote by $G(x, A)$ the expected number of visits by the Markov chain X before τ_A if $X_0 = x$. Then, it is a simple exercise to check that for all $n \geq 1$, $\mathbf{w} = (w_0, \ldots, w_n)$ with $w_0 = x$, $w_n \in A$ and $w_1, \ldots, w_{n-1} \in \mathcal{S} \setminus A$,

$$\mathbf{P}[L^A = \mathbf{w}] = \sum_{\mathbf{x}\,:\,L(\mathbf{x})=\mathbf{w}} \mathbf{P}[X[0, \tau_A] = \mathbf{x}]$$

$$= G(w_0, A)p(w_0, w_1)G(w_1, A \cup \{w_0\})p(w_1, w_2) \cdots$$
$$\times G(w_{n-1}, A \cup \{w_0, w_1, \ldots, w_{n-2}\})p(w_{n-1}, w_n).$$

It is therefore natural to define the function

$$F(w_0, \ldots, w_{n-1}; A) = \prod_{j=0}^{n-1} G(w_j, A \cup \{w_0, \ldots, w_{j-1}\}).$$

Again, it is a simple exercise on Markov chains to check that for all A', y and y',

$$G(y, A')G(y', A' \cup \{y\}) = G(y', A')G(y, A' \cup \{y'\}).$$

It follows immediately that F is in fact a symmetric function of its arguments. Hence,

$$\mathbf{P}[L_0^A = w_0, \ldots, L_\sigma^A = w_n | L_\sigma = w_n, L_{\sigma-1} = w_{n-1}]$$
$$= \frac{p(w_{n-1}, w_n)G(w_{n-1}, A)}{\mathbf{P}[L_\sigma = w_n, L_{\sigma-1} = w_{n-1}]}$$
$$\times \prod_{j=0}^{n-2} p(w_j, w_{j+1})G(w_j, (A \cup \{w_{n-1}\}) \cup \{w_0, \ldots, w_{j-1}\}).$$

This readily implies the Lemma when $j = 1$. Iterating this j times shows the Lemma. □

This Lemma shows that it is in fact fairly natural to index the loop-erased path backwards (define $\gamma_j = L_{\sigma-j}^A$, so that γ starts on A and goes back to $\gamma_\sigma = x$). Then, the time-reversal of loop-erased (conditioned and stopped) Markov chains have themselves a Markovian-type property.

Let us now come back to our two-dimensional setting: Suppose that X is a simple random walk on the grid $\delta\mathbb{Z}^2$ (we will then let the mesh δ of the lattice go to 0) that is started from 0. Let D denote some simply connected domain D with $0 \in D$ and $D \neq \mathbb{C}$, and let $D_\delta = \delta\mathbb{Z}^2 \cap D$, $A = A_\delta = \delta\mathbb{Z}^2 \setminus D$. We are interested in the behaviour when $\delta \to 0$ of the law of γ^δ which is defined as before as the time-reversed loop-erasure of $X[0, \tau_A]$. We now think on a heuristic level: First, note that the law of X_{τ_A} converges to the harmonic measure on ∂D from 0, so that it is possible to study the behaviour of γ^δ conditional on the value of $\{\gamma^\delta = y_0^\delta\}$ where $y_0^\delta \to y \in \partial D$ as $\delta \to 0$. Second, one might argue that on the one hand, simple random walk converges to planar Brownian motion which is conformally invariant, and that on the other hand the chronological loop-erasing procedure is purely geometrical to conclude

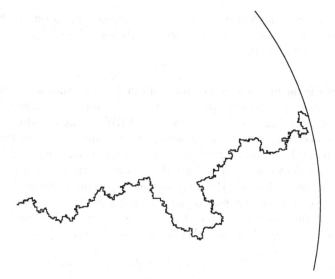

Fig. 1.3. A sample of the loop-erased random walk.

that when $\delta \to 0$, the law of γ^δ should converge to a conformal invariant curve that should be the loop-erasure of planar Brownian motion.

Unfortunately (or fortunately!), the geometry of planar Brownian curves is very complicated: It has points of any (even infinite) multiplicity (see e.g. [100]), loops at any scale, so that there is no "first" loop to erase, and decisions about what small microscopic loops to erase first may propagate to the decisions about what macroscopic loops one should erase. In other words, there is no simple (even random) algorithm to loop-erase a Brownian path in chronological order. Yet, the previous heuristic strongly suggests the law of γ^δ should converge, and that the limiting law is invariant under conformal transformations: The scaling limit of LERW in D should be (modulo time-change) identical to the conformal image of the scaling limit of LERW in D'. Furthermore, Lemma 1.1 should still be valid in the scaling limit. We now show that the combinations of these two properties in fact greatly reduce the family of possible scaling limits for LERW.

1.3 Iterations of Conformal Maps and SLE

We are therefore looking for the law of a random continuous curve $(\gamma_t, t \geq 0)$ with no self-crossings in the unit disc $\mathbb{U}$, with $\gamma_0 = 1$, $\lim_{t \to \infty} \gamma_t = 0$ that could be the scaling limit of (time-reversed) loop-erased random walk on a grid approximation of $\mathbb{U}$ (conditioned to exit $\mathbb{U}$ near 1). Define for each $t \geq 0$, the conformal map f_t from $\mathbb{U} \setminus \gamma[0, t]$ onto $\mathbb{U}$ which is normalized by $f_t(0) = 0$ and $f_t(\gamma_t) = 1$ (actually, if γ would have double-points, the domain of definition would be the connected component of $\mathbb{U} \setminus \gamma[0, t]$ that contains the origin, but let us a priori assume for convenience that γ is a simple curve).

It is easy to check that $t \mapsto |f'_t(0)|$ is an increasing continuous function that goes to ∞ as $t \to \infty$ (see for instance [2]). Hence, it is possible to reparametrize γ in such a way that

$$|f'_t(0)| = e^t. \tag{1.1}$$

This is the natural parametrization in our context. Indeed, let us now study the conditional law of $\gamma[t, \infty)$ given $\gamma[0, t]$. Lemma 1.1 suggests that this law is the scaling limit of (time-reversed) LERW in the slit domain $\mathbb{U} \setminus \gamma[0, t]$ conditioned to exit at γ_t, and conformal invariance then says that this is the same (modulo time-reparametrization) as the image under $z \mapsto f_t^{-1}(z)$ of an independent copy $\tilde{\gamma}$ of γ. Note that if one composes conformal maps that preserve the origin, then the derivative at the origin multiply: This shows that in fact, no time-change is necessary if we parametrize γ (and $\tilde{\gamma}$) by (1.1), in order for the conditional law of $(\gamma_{t+s}, s \geq 0)$ given $\gamma[0, t]$ to be identical to that of $(f_t^{-1}(\tilde{\gamma}_s), s \geq 0)$. In other words, for all fixed $t \geq 0$,

$$(f_{t+s}, s \geq 0) = (\tilde{f}_s \circ f_t, \ s \geq 0) \quad \text{in law}$$

where $(\tilde{f}_s, s \geq 0)$ is an independent copy of $(f_s, s \geq 0)$. In particular, $f_{2t} = \tilde{f}_t \circ f_t$ in law. Repeating this procedure, we see that for all $t \geq 0$ and all integer $n \geq 1$, f_{nt} is the iteration of n independent copies of f_t, and that f_t itself can be viewed as the iteration of n independent copies of $f_{t/n}$. In other words, $(f_t, t \geq 0)$ is an "infinitely divisible" process of conformal maps, and f_t is obtained by iterating infinitely many independent conformal maps that are infinitesimally close to the identity.

Back in the 1920's, Loewner observed that if $\gamma[0, \infty)$ is a simple continuous curve starting from 1 in the unit disc, then it is naturally encoded via a continuous function ζ_t taking its values on the unit circle. Let us now describe briefly how it goes. Suppose, as in the previous section, that $\gamma(0) = 1$, $\lim_{t \to \infty} \gamma_t = 0$ and that γ is parametrized in such a way that the modulus of the derivative at 0 of the conformal map f_t from $U_t := \mathbb{U} \setminus \gamma[0, t]$ into $\mathbb{U}$ that preserves the origin is e^t. Define $\zeta_t = (f'_t(0)/|f'_t(0)|)^{-1}$. In other words, if g_t denotes the conformal map from U_t onto $\mathbb{U}$ such that $g_t(0) = 0$ and $g'_t(0) = e^t \in (0, \infty)$, then

$$\zeta_t = g_t(\gamma_t)$$

and $g_t(z) = \zeta_t f_t(z)$. One can note (see e.g. [2, 49]) that for all $z \notin \gamma[0, t]$,

$$\partial_t g_t(z) = -g_t(z) \frac{g_t(z) + \zeta_t}{g_t(z) - \zeta_t}. \tag{1.2}$$

Hence, it is possible to recover γ from ζ as follows: For all $z \in \mathbb{U}$, define $g_t(z)$ as the unique solution to (1.2) starting from z. In case $g_t(z) = \zeta_t$ for some time t, then define $\gamma_t = g_t^{-1}(\zeta_t)$ (we know already a priori that since γ is a simple curve, the map g_t^{-1} extends continuously to the boundary). Note that if $g_t(z) = \zeta_t$, then $g_s(z)$ is not well-defined for $s \geq t$.

Hence, in order to define the random curve γ that should be the scaling limit of loop-erased random walks, it suffices to define the random function $\zeta_t = \exp(iW_t)$, where $(W_t, t \geq 0)$ is real-valued. Our previous considerations suggest that the following conditions should be satisfied:

- The process W is almost surely continuous.
- The process W has stationary increments (this is because g_t is obtained by iterations of identically distributed conformal maps)
- The laws of the processes W and $-W$ are identical (this is because the law of L and the law of the complex conjugate $\overline{L}$ are identical).

The theory of Markov processes tells us that the only possible choices are: $W_t = \beta_{\kappa t}$ where β is standard Brownian motion and $\kappa \geq 0$ a fixed constant. In order to simplify some future notations, we will usually write

$$W_t = \sqrt{\kappa}B_t, t \geq 0$$

where $(B_t, t \geq 0)$ is standard (one-dimensional) Brownian motion.

In summary, we have just seen that on a heuristic level, if the scaling limit of loop-erased random walk exists and is conformally invariant, then the scaling limit in the unit disk should be described as follows: For some fixed constant $\kappa = \kappa_{LERW}$, define $\zeta_t = \exp(i\sqrt{\kappa}B_t), t \geq 0$, solve for each $z \in \mathbb{U}$, the equation (1.2) with $g_0(z) = z$. This defines a conformal map g_t from the subset U_t of the unit disk onto $\mathbb{U}$. Then, one can construct γ because

$$U_t = \mathbb{U} \setminus \gamma[0, t]$$

and

$$\gamma_t = g_t^{-1}(\zeta_t).$$

As we shall see later on in the lectures, this heuristic arguments can be made rigorous, and it will turn out that $\kappa_{LERW} = 2$.

1.4 The Critical Percolation Exploration Process

In the context of LERW, the random curve joins a point in the inside of the domain to a point on the boundary of the domain. In statistical physics models, one is often interested in "interfaces". Some of these interfaces appear to be random curves from one point on the boundary to another point on the boundary. A natural setup is to study curves from 0 to infinity in the upper half-plane $\mathbb{H} := \{x + iy : y > 0\}$. Then, we look for random non-self-crossing curves γ such that the law of $\gamma[t, \infty)$ given $\gamma[0, t]$ has the same law than the conformal image of an independent copy $\tilde{\gamma}$ of γ under a conformal map from $\mathbb{H}$ onto $\mathbb{H} \setminus \gamma[0, t]$ that maps ∞ onto itself and 0 onto γ_t.

We now very briefly describe an important discrete model for which it has now also been proved that it behaves in a conformally invariant way in the scaling limit (more details on the model and its conformal covariance will be given in Chapter 10): Critical site percolation on the triangular lattice. Actually, it is more convenient to describe this in terms of cell-colouring of the honeycombe lattice. Suppose that a simply connected domain D is fixed, as well as two distinct points a and b on ∂D. Let D_δ denote a suitably chosen approximation of D by a simply connected union of hexagonal cells of size δ. Let a_δ (resp. b_δ) denote a vertex of the honeycombe lattice on ∂D_δ that is close to a (resp. to b). Then, the cells on ∂D_δ can be divided into two "arcs" B_δ and W_δ in such a way that a_δ, B_δ, b_δ and W_δ are oriented clockwise "around" D_δ. Decide that all hexagons in B_δ are colored in black and that all hexagons in W_δ are colored in white. On the other hand, all other cells in D_δ are chosen to be black or white with probability $1/2$ independently of each other. Consider now the (random) path γ_δ from a_δ to b_δ that separates the cluster of black hexagons containing B_δ from the cluster of white hexagons containing W_δ.

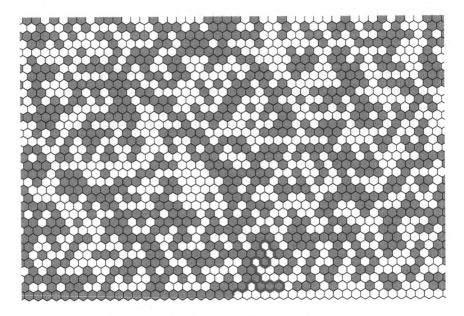

Fig. 1.4. The beginning of the discrete exploration process.

For deep reasons that will be discussed later in these lectures, it will turn out that when $\delta \to 0$, the law of γ_δ converges towards that of a random curve γ from a to b in D, and that the law of that curve is conformally invariant: The law of $\Phi(\gamma)$ when Φ is a conformal map from D onto $\Phi(D)$ is that of the corresponding path (i.e. of the scaling limit of percolation cluster interfaces) from $\Phi(a)$ to $\Phi(b)$ in $\Phi(D)$.

Again, on the discrete level, it is easy to see that γ_δ has the same type of Markovian property that LERW. More precisely, conditioning on the first steps of γ is equivalent to condition the percolation process to have black hexagons on the left-boundary of these steps and white hexagons on the right side. Hence, the conditional law of the remaining steps is that of the percolation interface in the new domain obtained by slitting D_δ along the first steps of γ_δ. Figure 1.4 shows the beginning of the interface γ_δ in the case where D is the upper half-plane.

Another equivalent way to define the interface γ_δ goes as follows: It is a myopic self-avoiding walk. At each step γ_δ looks at its three neighbours (on the honeycomb lattice) and chooses at random one of the sites that it has not visited yet (there are one or two such sites since one site is anyway forbidden because it was the previous location of the walk).

This discrete walk in the upper half-plane is a very special discrete model that will turn out to converge to an SLE. The corresponding value of κ is 6. Here, the starting point $a = 0$ and the end-point $b = \infty$ are both on the boundary of the domain, so that the previous definition of radial SLE is not well-suited anymore.

Fig. 1.5. The exploration process, proved to converge to SLE_6 (see Chapter 10)

1.5 Chordal versus Radial

The natural time-parametrization in the previous setup goes as follows: Let g_t denote the conformal map from $\mathbb{H} \setminus \gamma[0, t]$ onto $\mathbb{H}$ that is normalized at infinity in the sense that when $z \to \infty$,

$$g_t(z) = z + \frac{a_t}{z} + o(1/z).$$

It is easy to see that a_t is positive, increasing and that it is natural to parametrize g_t in such a way that a_t is a multiple of t (since the a_t terms add up when one composes two such conformal maps). It is natural to choose $a_t = 2t$ (this is consistent with the chosen parametrization in the radial case). Then, define $w_t = g_t(\gamma_t)$, and observe that

$$\partial_t g_t(z) = \frac{2}{g_t(z) - w_t}. \tag{1.3}$$

Hence, just as in the radial case, we observe that it is possible to recover γ using w, and that the only choice for w that is consistent with the "Markovian property" is to take $w_t = \sqrt{\kappa} B_t$, where B is ordinary one-dimensional Brownian motion.

Hence, one is lead to the following definition: Let $w_t = \sqrt{\kappa} B_t$, and define for all $z \in \overline{\mathbb{H}}$, the solution $g_t(z)$ of (1.3) up to the (possibly infinite) time $T(z)$ at which $g_t(z)$ hits w_t. Then, define

$$H_t = \{z \in \mathbb{H} \ : \ T(z) > t\}$$

and

$$K_t = \{z \in \overline{\mathbb{H}} \ : \ T(z) \le t\}.$$

Then, g_t is the normalized conformal map from H_t onto $\mathbb{H}$. We call $(K_t, t \ge 0)$ the chordal SLE_κ in the upper half-plane.

It turns out that radial and chordal SLE's are rather closely related: Consider for instance, the conformal image of radial SLE_κ under the map that maps $\mathbb{U}$ onto $\mathbb{H}$, 1 to 0 and 0 to i. Consider both this process and chordal SLE_κ up to their first hitting of the circle of radius $1/2$ around zero say. Then, the laws of these two processes are absolutely continuous with respect to each other [87]. This justifies a posteriori the choice of time-parametrization in the chordal case.

1.6 Conclusion

We have seen that if one considers a discrete model of random curves (or interfaces) that combine the two important features:

- The Markovian type property in the discrete setting,
- Conformal invariance in the limit when the mesh of the lattice goes to zero,

then the good way to construct the possible candidates for the scaling limit of these curves is to encode them via the corresponding conformal mappings. Then, these (random) conformal mappings are themselves obtained by iterations of identically distributed random conformal maps. Loewner's theory shows that such families of conformal maps are themselves encoded by a one-dimensional function. If one knows this one-dimensional function, one can recover the family of conformal maps, and therefore also the two-dimensional curve. The one-dimensional random function that generates the scaling limits of the discrete models must necessarily be a one-dimensional Brownian motion. The corresponding random two-dimensional curves are SLE processes.

Bibliographical Comments

Most of the intuition about how to define radial and chordal SLE (with LERW as a guide) was already present in the introduction of Oded Schramm's first paper [123] on SLE that he released in March 1999. Our presentation of Lemma 1.1 is borrowed from Lawler [81], but there are other proofs of it (it is for instance closely related to Wilson's algorithm [136] that will be discussed in Chapter 9).

2 Loewner Chains

This chapter does contain background material on conformal maps and on Loewner's equation (no really new results will be presented here). The setup is deterministic in this Chapter. SLE will be introduced in the next Chapter.

2.1 Measuring the Size of Subsets of the Half-Plane

We study increasing "continuously growing" compact subsets $(K_t, t \geq 0)$ of the upper half-plane. It will turn out to be important to choose the good time-parametrization. We want to find the natural way to measure the size $a(K)$ of a compact set K and we will then choose the time-parametrization in such a way that $a(K_t) = t$. We will use the following definition throughout the paper.

Definition. We say that a compact subset K of the closed upper half-plane $\overline{\mathbb{H}}$, such that $H := \mathbb{H} \setminus K$ is simply connected, is a hull.

Riemann's mapping theorem asserts that there exist conformal maps Φ from H onto $\mathbb{H}$ with $\Phi(\infty) = \infty$. Actually, if Φ is such a map, the family of maps $b\Phi + b'$ for real b' and positive b is exactly the family of conformal maps from H onto $\mathbb{H}$ that fix infinity.

Note that since K is compact, the mapping $\Psi : z \mapsto 1/\Phi(1/z)$ is well-defined on a neighbourhood of 0 in $\mathbb{H}$. It is possible to extend this map Ψ to a whole neighbourhood of 0 in the plane by reflection along the real axis (this is usually called Schwarz reflection) and to check that this extension is analytic. This implies that Φ can be expanded near infinity: There exist $b_1, b_0, b_{-1}, \ldots$, such that

$$\Phi(z) = b_1 z + b_0 + b_{-1} z^{-1} + \cdots + b_{-n} z^{-n} + o(z^{-n})$$

when $z \to \infty$ in $\mathbb{H}$. Furthermore, since Φ preserves the real axis near infinity, all coefficients b_j are real.

Hence, for each K, there exists a unique conformal map $\Phi = \Phi_K$ from $H = \mathbb{H} \setminus K$ onto $\mathbb{H}$ such that:

$$\Phi(z) = z + 0 + o(1/z) \text{ when } z \to \infty.$$

This is sometimes called the hydrodynamical normalization. In particular, there exists a real $a = a(K)$ such that

$$\Phi(z) = z + \frac{2a}{z} + o(1/z) \text{ when } z \to \infty.$$

This number $a(K)$ is a way to measure the size of K. In a way, it tells "how big K is in $\mathbb{H}$, seen from infinity". It may a priori not be clear that a is a non-negative increasing function of the set K. There is a simple probabilistic interpretation of $a(K)$ that immediately implies these facts: Suppose that $Z = X + iY$ is a complex Brownian motion started from $Z_0 = iy$ (for some large y, so that $Z_0 \notin K$) and stopped at its first exit time τ of H. The expansion $\Phi(z) = z + o(1)$ near infinity shows that $\Im(\Phi(z) - z)$ is a bounded harmonic function in H. The martingale stopping theorem therefore shows that

$$\mathbf{E}[\Im(\Phi(Z_\tau)) - Y_\tau] = \Im(\Phi(iy) - iy) = \frac{2a}{iy} + o(1/y).$$

But $\Phi(Z_\tau)$ is real because of the definition of τ. Therefore

$$2a = \lim_{y \to +\infty} y\, \mathbf{E}[\Im(Y_\tau)].$$

In particular, $a \geq 0$.

One can also view a as a function of the normalized conformal map Φ_K instead of K. The chain rule for Taylor expansions then immediately shows that

$$a(\Phi^1 \circ \Phi^2) = a(\Phi^1) + a(\Phi^2)$$

for any two normalized maps Φ^1 and Φ^2. In particular, this readily implies that $a(K) \leq a(K')$ if $K \subset K'$ (because there exists a normalized conformal map from $\mathbb{H} \setminus \Phi_K(K' \setminus K)$ onto $\mathbb{H}$).

Let us now observe two simple facts:

- If $\lambda > 0$, then $a(\lambda K) = \lambda^2 a(K)$. This is simply due to the fact that

$$\Phi(z/\lambda) = \frac{z}{\lambda} + \frac{2a(K)\lambda}{z} + o(\lambda/z)$$

so that

$$\Phi_{\lambda K}(z) = \lambda \Phi_K(z/\lambda) = z + \frac{2a(K)\lambda^2}{z} + o\lambda(\lambda/z) \qquad (2.1)$$

when $z \to \infty$.

- When K is the vertical slit $[0, iy]$, then

$$\Phi_K(z) = \sqrt{z^2 + y^2}.$$

In particular, we see that $a([0, iy]) = y^2/4$. Note that if y is very small, the actual diameter of the vertical slit $[0, iy]$ is much larger than $a([0, iy])$.

Equation (2.1) shows that for all K such that $a(K) = 1$, one has $a(\sqrt{\lambda}K) = \lambda$ and

$$\lim_{\lambda \to 0} \frac{\Phi_{\sqrt{\lambda}K}(z) - \Phi_{\{0\}}(z)}{\lambda} = \lim_{\lambda \to 0} \frac{\Phi_{\sqrt{\lambda}K}(z) - z}{\lambda} = \frac{2}{z}. \qquad (2.2)$$

Actually, it is not very difficult to prove that for all given r, there exists $C > 0$ such that this convergence takes place uniformly over all K of radius smaller than r and $|z| > Cr$. See Lemma 2.7 in [86].

2.2 Loewner Chains

Suppose that a continuous real function w_t with $w_0 = 0$ is given. For each $z \in \overline{\mathbb{H}}$, define the function $g_t(z)$ as the solution to the ODE

$$\partial_t g_t(z) = \frac{2}{g_t(z) - w_t} \tag{2.3}$$

with $g_0(z) = z$. This is well-defined as long as $g_t(z) - w_t$ does not hit 0, i.e., for all $t < T(z)$, where

$$T(z) := \sup\{t \geq 0 \ : \ \min_{s \in [0,t]} |g_s(z) - w_s| > 0\}.$$

We define

$$K_t := \{z \in \overline{\mathbb{H}} \ : \ T(z) \leq t\}$$
$$H_t := \mathbb{H} \setminus K_t.$$

Note for instance that if $w_t = 0$ for all t, then

$$g_t(z) = \sqrt{z^2 + 4t}$$

and $K_t = [0, 2i\sqrt{t}]$.

It is very easy to check that g_t is a bijection from H_t onto $\mathbb{H}$ (in order to see that it is surjective, one can just look at the ODE "backwards in time" to find which point z is such that $g_t(z) = y$). Moreover K_t is bounded (because w is continuous and bounded on $[0, t]$) and H_t has a unique connected component (because g_t^{-1} is continuous). Standard arguments from the theory of ordinary differential equations can be applied to check that g_t is analytic and that one can formally differentiate the ODE with respect to z, so that

$$\partial_t g_t'(z) = \frac{-2g_t'(z)}{(g_t(z) - w_t)^2}.$$

So, g_t is a conformal map from H_t onto $\mathbb{H}$.

Note also that $|\partial_t g_t(z)|$ is uniformly bounded when z is large and t belongs to a given finite interval $[0, t_0]$. Hence, it follows that $g_t(z) = z + O(1)$ near infinity and uniformly over $t \in [0, t_0]$. Hence (using the ODE yet again), $\partial_t g_t(z) = 2/z + o(1/z)$ uniformly over $t \in [0, t_0]$ so that finally, for each t,

$$g_t(z) = z + \frac{2t}{z} + o(1/z)$$

when $z \to \infty$. In other words, $a(K_t) = t$. The family $(K_t, t \geq 0)$ is called the Loewner chain associated to the driving function $(w_t, t \geq 0)$.

Loewner's original motivation was to control the behaviour of the coefficients of the Taylor expansion of conformal maps and for this goal, it is sufficient to consider smooth slit domains (see e.g., [2, 49]). For this reason, the following question was only addressed later (see [72]): If the continuous function $(w_t, t \geq 0)$ is given, what can be said about the family of compact sets $(K_t, t \geq 0)$?

In the introduction, we started with a continuous curve γ, then using γ, we defined H_t, the conformal maps g_t, the function w_t and argued that one could recover γ from w_t, using the fact that we a priori knew that g_t^{-1} extends continuously to $w_t \in \partial \mathbb{H}$ and that $g_t^{-1}(w_t)$ was well-defined (and equal to γ_t) because γ is a continuous curve. But if one starts with a general continuous function w_t, then it can in fact happen that g_t^{-1} does not extend continuously to w_t.

Before making general considerations, let us exhibit a simple example to show that $(K_t, t \geq 0)$ does not need to be a simple curve. For $\theta \in [0, \pi)$, let $\eta(\theta) = \exp(i\theta) - 1$. Define $t(\theta) = a(\eta[0, \theta])$ the "size" of the arc $\eta[0, \theta]$. Finally, define the reparametrization γ of η in such a way that $a(\gamma[0, t]) = t$. γ is defined for all $t < T := \lim_{\theta \to \pi-} a(\eta[0, \theta])$. It is simple to see that there exists a continuous function $(w_t, t < T)$ such that the normalized conformal maps g_t from $\mathbb{H} \setminus \gamma[0, t]$ onto $\mathbb{H}$ satisfy the equation (2.3). Furthermore, when $t \to T-$, w_t converges to a finite limit w_T. At time T, the curve $\gamma[0, T]$ disconnects the inside of the semi-circle from the outside. Just before T, because g_t is normalized "from infinity", the inside of the semi-circle is mapped onto a small region which is very close to $w_t = g_t(\gamma_t)$. When $t \to T-$, all points inside the semi-circle are hitting w_T. In other words, K_T is the whole semi-disc, H_T is the complement of the semi-disc, and g_T is the normalized map from the simply connected domain H_T onto $\mathbb{H}$.

Let us now give a couple of general definitions:

- We say that $(K_t, t \geq 0)$ is a simple curve if there exists a simple continuous curve γ such that $K_t = \gamma[0, t]$.
- We say that $(K_t, t \geq 0)$ is generated by a curve if there exists a continuous curve γ with no self-crossings, such that for all $t \geq 0$, $H_t = \mathbb{H} \setminus K_t$ is the unbounded connected component of $\mathbb{H} \setminus \gamma[0, t]$. In other words, K_t is the union of γ and of the inside of the loops that γ creates.
- We say that $(K_t, t \geq 0)$ is pathological if it is not generated by a curve.

In each of these three cases, one can find (deterministic) continuous functions w_t such that the family $(K_t, t \geq 0)$ that it constructs falls into this category: For the first case, consider for instance $w_t = 0$ as before, for the second case, one can use the example with the semi-circle. For the more intricate third case, let us mention the following example (due to Don Marshal and Steffen Rohde, see [108]): Let γ denote a simple curve in $\mathbb{H}$ started from $\gamma_0 = 0$ that spirals clockwise around the segment $[i, 2i]$ an infinite number of times, and then unwinds itself. Then at the "time" at which it winds around the segment an infinite number of times, γ is not continuous i.e. $K_t \setminus K_{t-}$ is the whole segment. However, this Loewner chain corresponds to a continuous function w_t. Such pathologies could arise at any scale.

We now characterize the families $(K_t, t \geq 0)$ of compact sets that are Loewner chains:

Proposition 2.1 *The following two conditions are equivalent:*

1. $(K_t, t \geq 0)$ *is a Loewner chain associated to a continuous driving function* $(w_t, t \geq 0)$.
2. *For all* $t \geq 0$, $a(K_t) = t$, *and for all* $T > 0$, *and* $\varepsilon > 0$, *there exists* $\delta > 0$ *such that for all* $t \leq T$, *there exists a bounded connected set* $S \subset \mathbb{H} \setminus K_t$ *with diameter not larger than* ε *such that* S *disconnects* $K_{t+\delta} \setminus K_t$ *from infinity in* $\mathbb{H} \setminus K_t$.

Sketch of the proof. Let us now prove that 2. implies 1. (the fact that 1. implies 2. is very easy): 2. implies that for all $t \geq 0$, the diameter of the sets $g_t(K_{t+\delta} \setminus K_t)$ decrease towards 0 when $\delta \to 0$. Hence, one can simply define w_t by

$$\{w_t\} = \lim_{\delta \to 0} \overline{g_t(K_{t+\delta} \setminus K_t)}.$$

Then, one uses 2. to show that $t \mapsto w_t$ is uniformly continuous. It then only remains to check that indeed

$$\lim_{\delta \to 0} \frac{g_{t+\delta}(z) - g_t(z)}{\delta} = \frac{2}{g_t(z) - w_t}.$$

This is achieved by applying the uniform version of (2.2).   □

Suppose now that K_t is the Loewner chain

$$K_t = [0, c\sqrt{t}]$$

for some $c = c(\theta) \exp(i\theta) \in \mathbb{H}$. Here, $\theta \neq 0$ is given, and then the positive real $c(\theta)$ is chosen in such a way that $a(K_1) = 1$. Scaling immediately shows that $a(K_t) = t$ for all $t > 0$, so that there exists therefore a continuous driving function w that generates these slits. Again, scaling (because $K_{\lambda t} = \sqrt{\lambda} K_t$) shows that necessarily, this function w must be of the type

$$w_t = c_1 \sqrt{t}$$

for some real constant $c_1 = c_1(\theta)$. Let g_t^θ denote the corresponding family of conformal maps.

Let us now choose a new driving function w as follows: $w_t = 0$ when $t < 1$ and for $t \geq 1$:

$$w_t = c_1 \sqrt{t - 1}.$$

When $t < 1$, then K_t is just the straight slit. In particular, $g_1(z) = \sqrt{z^2 + 4}$. When $t > 1$, then K_t is obtained by mapping the angled slit K_{t-1}^θ back by g_1^{-1}. In particular, we see that the curve γ generated by this function w is not differentiable at $t = 1$. This is one simple hint to the fact that Hölder-1/2 regularity may be critical (note that at $t = 1$, w is just Hölder 1/2).

The general relation between smoothness of the driving function and regularity of the slit has also recently been investigated (in the deterministic setting) by Marshall-Rohde [108]. In this paper, it is shown that Hölder-1/2 is in a sense a "critical regularity" for the driving function w_t: Loosely speaking (their results are more precise than that), if w is better than Hölder-1/2, then it defines a "smooth" (in some appropriate sense) slit, but nasty "pathological" phenomena can occur for Hölder-1/2 driving functions. See [108] and the references therein.

Bibliographical Comments

For general background on complex analysis, Riemann's mapping theorem, there are plenty of good references, see for instance [1, 119]. Loewner introduced his equation (in the radial setting) in 1923 [103]. For general information about Loewner's equation, and in particular how Loewner used it to prove that $|a_3| \leq 3$ for univalent functions $z + \sum_{n \geq 2} a_n z^n$ on $\mathbb{U}$ as well as other applications, see for instance [2, 49]. For how it is used in de Branges' proof of the Bieberbach conjecture, a good self-contained reference is Hayman's book [58]. For basics on hypergeometric functions, see e.g., [99].

Proposition 2.1 is derived in [86], see also [114]. Carleson and Makarov [35, 36] have used Loewner's (radial) equation in the context of Diffusion Limited Aggregation.

3 Chordal SLE

3.1 Definition

Chordal SLE_κ is the Loewner chain $(K_t, t \geq 0)$ that is obtained when the driving function

$$w_t = W_t := \sqrt{\kappa} B_t$$

is $\sqrt{\kappa}$ times a standard real-valued Brownian motion $(B_t, t \geq 0)$ with $B_0 = 0$. Let us now list a couple of consequences of the simple properties of Brownian motion:

- Brownian motion is a strong Markov process with independent increments. This implies that for any stopping time T (with respect to the natural filtration $(\mathcal{F}_t, t \geq 0)$ of B), the process

$$(g_{T+t}(K_{T+t} \setminus K_T) - W_T, t \geq 0)$$

is independent of $\mathcal{F}_T$ and that its law is identical to that of $(K_t, t \geq 0)$. Note that one has to shift by W_T in order to obtain a process starting at the origin.

- Brownian motion is scale-invariant: For each $\lambda > 0$, the process $W_t^\lambda := W_{\lambda t}/\sqrt{\lambda}, t \geq 0$ has the same law than W. But

$$\partial_t(g_{\lambda t}(\sqrt{\lambda} z)) = \frac{2\lambda}{g_t(\sqrt{\lambda} z) - W_{\lambda t}}.$$

In particular, if

$$g_t^\lambda(z) := g_{\lambda t}(z\sqrt{\lambda})/\sqrt{\lambda},$$

then

$$\partial_t g_t^\lambda(z) = \frac{2}{g_t^\lambda(z) - W_t^\lambda}$$

and $g_0^\lambda(z) = z$. In other words, $(K_{\lambda t}, t \geq 0)$ and $(\sqrt{\lambda} K_t, t \geq 0)$ have the same law: Chordal SLE_κ is scale-invariant.

- Brownian motion is symmetric (W and $-W$ have the same law). Hence, the law of $(K_t, t \geq 0)$ is symmetric with respect to the imaginary axis.

It is actually possible to prove the following result:

Proposition 3.1 *For all $\kappa \geq 0$, chordal SLE_κ is almost surely not pathological. When $\kappa \leq 4$, it is a.s. a simple curve γ, when $\kappa > 4$, it is a.s. generated by a (non-simple) curve γ.*

This result is due to Rohde-Schramm [118] (see [93] for the critical case $\kappa = 8$). It is not an easy result, especially for the values $\kappa > 4$. Actually, while this fact is important and useful in order to understand heuristically the behaviour and the properties of SLE_κ, it turns out that one can derive many of them without knowing that the SLE_κ is generated by a continuous curve. We therefore omit the proof in these lectures, and we will call $(K_t, t \geq 0)$ the SLE. In some cases that we will focus on ($\kappa = 2, 8/3, 6, 8$), the fact that SLE_κ is a.s. generated by a curve will actually follow from other considerations.

It is however easy to see that $\kappa = 4$ is a critical value: Consider chordal SLE_κ, and define

$$X_t = \frac{g_t(1) - W_t}{\sqrt{\kappa}}.$$

Note that X hits zero if and only if the chordal SLE absorbs the boundary point 1. But X satisfies

$$dX_t = dB_t + \frac{2}{\kappa X_t}dt. \tag{3.1}$$

It is a $1+(4/\kappa)$ dimensional Bessel process, and it is well-known (see e.g. [117]) that such a process a.s. hits zero if and only if $\kappa > 4$. This can for instance be viewed as a consequence of the fact that if X is a Bessel process of dimension d started away from zero, then if $d \neq 2$, X^{2-d} is a local martingale, and when $d = 2$, $\log X$ is a local martingale.

It follows that:

Proposition 3.2 • *If $\kappa \leq 4$, then almost surely $\cup_{t \geq 0} K_t \cap \mathbb{R} = \{0\}$.*
• *If $\kappa > 4$, then almost surely, $\mathbb{R} \subset \cup_{t \geq 0} K_t$.*

Assuming that the SLE is generated by a curve, this readily shows that the SLE curve is simple if and only if $\kappa \leq 4$.

If one defines, for all $z \in \mathbb{H}$, the solution X_t^z to (3.1) started from $X_0^z = z/\sqrt{\kappa}$ (up to the stopping time $T(z)$). Then, we see that SLE_κ can be interpreted in terms of the flow of a complex Bessel process: For each $t > 0$, K_t is the set of starting points such that X_t^z has hit 0 before time t.

3.2 A First Computation

We now compute the probability of some simple events involving the chordal Schramm-Loewner evolution. Suppose that $a < 0 < c$. Let $\kappa > 0$ be fixed. Define the event $E_{a,c}$ that the chordal SLE_κ hits $[c, \infty)$ before $(-\infty, a]$. For the reasons that we just discussed, this makes sense only if $\kappa > 4$ (otherwise, it never hits these intervals). The goal of this section is to compute the probability of $E_{a,c}$. The scaling property of chordal SLE shows that this is a function of the ration c/a only. We can therefore define $F = F_\kappa$ on the interval $(0, 1)$ by

$$\mathbf{P}[E_{a,c}] = F(-a/(c - a)).$$

Proposition 3.3 *For all $\kappa > 4$ and $z \in (0, 1)$,*

$$F(z) = c(\kappa) \int_0^z \frac{du}{u^{4/\kappa}(1 - u)^{4/\kappa}}$$

where $c(\kappa) = (\int_0^1 u^{-4/\kappa}(1 - u)^{-4/\kappa}du)^{-1}$ is chosen so that $F(1) = 1$.

Note that this Proposition is in fact a property of the real Bessel flow: $E_{a,c}$ is the event that X^c hits 0 before X^a does.

Proof. Suppose that $\mathcal{F}_t = \sigma(B_s, s \le t)$ is the natural filtration associated to the Brownian motion, and define $T_a = T(a)$ and $T_c = T(c)$ as before (the times at which a and c are respectively absorbed by K_t). For $t < T_a$ and $t < T_c$ respectively, define

$$A_t := g_t(a) \text{ and } C_t := g_t(c).$$

Suppose that $t < \min(T_a, T_c)$, and define

$$K_{t,s} = g_t(K_{t+s} \setminus K_t) - W_t.$$

The strong Markov property shows that $(K_{t,s}, s \ge 0)$ is also chordal SLE_κ, and that it is independent from $\mathcal{F}_t$. Also, if $t < \min(T_a, T_c)$, the event $E_{a,c}$ corresponds to the event that $(K_{t,s}, s \ge 0)$ hits $[C_t - W_t, \infty)$ before $(-\infty, A_t - W_t]$. Hence, if $t < \min(T_a, T_c)$,

$$\mathbf{P}[E_{a,c} \mid \mathcal{F}_t] = F\left(\frac{W_t - A_t}{C_t - A_t}\right).$$

In particular, this shows that the right-hand side of the previous identity is a (bounded) martingale. We know that $W_t = \sqrt{\kappa}B_t$, and that

$$\partial_t A_t = \frac{2}{A_t - W_t}, \quad \partial_t C_t = \frac{2}{C_t - W_t}.$$

Hence, if we put $Z_t := (W_t - A_t)/(C_t - A_t)$, stochastic calculus yields

$$dZ_t = \frac{\sqrt{\kappa}dB_t}{C_t - A_t} + \frac{2dt}{(C_t - A_t)^2}\left(\frac{1}{Z_t} - \frac{1}{1 - Z_t}\right).$$

One can now also introduce the natural time-change

$$s = s(t) := \int_0^t \frac{du}{(C_u - A_u)^2}$$

and define $\tilde{Z}$ in such a way that $\tilde{Z}_{s(t)} = Z_t$. Then,

$$\tilde{Z}_s = \sqrt{\kappa}d\tilde{B}_s + 2\left(\frac{1}{Z_s} - \frac{1}{1 - Z_s}\right)ds$$

where $(\tilde{B}_s, s \ge 0)$ is a standard Brownian motion.

But K_t hits $(-\infty, a)$ if and only if Z_t hits 0, and K_t hits (c, ∞) if and only if Z_t hits 1. Hence, $F(z)$ is the probability that the diffusion $\tilde{Z}$ started from $\tilde{Z}_0 = z$ hits 1 before 0. One can invoke (for instance) the general theory of diffusions to argue that the function F is therefore smooth on $(0,1)$. Itô's formula (since $F(\tilde{Z}_s)$ is a martingale) then implies that

$$\frac{\kappa}{4} F''(z) + \left(\frac{1}{z} - \frac{1}{1-z} \right) F'(z) = 0. \tag{3.2}$$

Furthermore, the boundary values of F are simple to work out: When $\kappa > 4$, one can see (for instance comparing $\tilde{Z}$ with a Bessel process) that

$$\lim_{z \to 0} F(z) = 0 \text{ and } \lim_{z \to 1} F(z) = 1.$$

Hence, F is the only solution to the ODE (3.2) with boundary values $F(0) = 0$ and $F(1) = 1$. This immediately proves the Proposition. $\qquad\square$

Note that when $z \to 0$,

$$F(z) \sim \frac{c(\kappa)}{1 - 4/\kappa} z^{1 - 4/\kappa}.$$

In particular, for $\kappa = 6$, we get the exponent $1/3$.

Exactly in the same way, it is possible (for $\kappa > 4$) to compute the probability that chordal SLE_κ (started from 0) hits the interval $[a, c]$ before $[c, \infty)$ when $0 < a < c$. This is a function $\tilde{F}$ of the ratio a/c, satisfying a linear second-order differential equation, with the boundary conditions

$$\tilde{F}(1) = 0 \text{ and } \tilde{F}(0) = 1.$$

3.3 Chordal SLE_κ in Other Domains

Suppose that D is some given non-empty open simply connected subset of the complex plane with $D \neq \mathbb{C}$. We do not impose any regularity condition on ∂D. Riemann's mapping theorem shows that there exist (many) conformal maps Φ from the upper half-plane $\mathbb{H}$ onto D. Even if the boundary of ∂D is not smooth, one can define a general notion that coincides with that of boundary points when it is smooth: For each $x \in \mathbb{R}$, we say that (if some map Φ is given) $\Phi(x)$ is a prime end of D (see e.g. [116] for a more precise and correct definition).

Suppose that O and U are two distinct prime ends in D. Then, there exists a conformal map Φ from $\mathbb{H}$ onto D such that $\Phi(0) = O$ and $\Phi(\infty) = U$. Actually, this only characterizes $\Phi(\cdot)$ up to a multiplicative factor (because $\Phi(\lambda \cdot)$ would then also do).

Suppose that $(K_t, t \geq 0)$ is chordal SLE_κ in $\mathbb{H}$ as defined before. We define SLE_κ in D from O to U as the image of the process $(K_t, t \geq 0)$ under Φ. Recall that Φ is defined up to a multiplicative constant. However, the scaling property of SLE_κ in $\mathbb{H}$ shows that the law of $(\Phi(K_t), t \geq 0)$ is invariant (modulo linear time-change) if we replace $\Phi(\cdot)$ by $\Phi(\lambda \cdot)$.

To illustrate this definition, consider the following setup: Suppose that $\kappa = 6$ and that OAC is an equilateral triangle. Let Φ denote the conformal map from $\mathbb{H}$ onto the triangle defined in such a way that

$$\Phi(a) = A, \Phi(0) = O, \Phi(c) = C$$

where $a < 0 < c$ are given. This conformal map can be easily described explicitly using the Schwarz-Christoffel transformations [1, 119]. Note that $U = \Phi(\infty)$ is on the interval AC. It turns out that

$$\frac{AU}{AC} = F(z)$$

where $z = -a/(c - a)$ and $F = F_{\kappa=6}$ is precisely the same hypergeometric function as in Proposition 3.3. Hence, the probability that chordal SLE_6 from O to U in the equilateral triangle OAC hits AU before UC is simply the ratio AU/AC.

Suppose now that $\kappa \in (4, 8)$. Just as for the hypergeometric function F, the functions $\tilde{F}$ that were defined at the end of the last subsection have a nice interpretation in terms of conformal mappings onto triangles: Consider an isocele triangle $\mathcal{T} = OAU$ with $OA = AU = 1$ and angle $\pi(1 - 4/\kappa)$ at the vertices O and U. The angle at the vertex A is therefore $\pi(8/\kappa - 1)$. Consider now a chordal SLE_κ from O to U in the triangle $\mathcal{T}$. Let X denote the random point at which it first hits the segment AU.

Proposition 3.4 *The law of X is the uniform distribution on AU.*

This is a direct consequence of the explicit computation of $\tilde{F}$ and of the explicit Schwarz-Christoffel mapping from the upper half-plane onto $\mathcal{T}$: For each $C \in AU$, one can compute the probability that $X \in [AC]$ via the function $\tilde{F}$.   □

This gives a first justification to the fact that the only possible conformally invariant scaling limit of the critical percolation exploration process is SLE_6 (see more on this in Chapter 10). Indeed, suppose that the critical percolation exploration process is conformally invariant. We have argued in the first chapter that the scaling limit is one of the SLEs. Suppose that it is SLE_κ for a given value of κ, and consider the corresponding triangle $\mathcal{T}$.

Clearly in the discrete case (for a fixed small meshsize), up to the first time at which it hits the edge AU, the critical exploration process from O to U and the critical exploration process from O to A in $\mathcal{T}$ coincide. Hence, the hitting distributions on AU for chordal SLE_κ from O to U and for chordal SLE_κ from O to A coincide. In particular, the uniform distribution on AU must be invariant under the anti-conformal map from $\mathcal{T}$ onto itself that maps O onto itself and interchanges the vertices A and U. This is only true when the triangle is symmetric (i.e. the angles at U and A are identical), in other words when $\alpha = \pi/3$ or $\kappa = 6$.

We shall see in the next chapter that indeed, for SLE_6, the whole paths from O to A and from O to U coincide up to their first hitting of AU. This is the so-called locality property of SLE_6.

3.4 Transience

We conclude this chapter with the following fact (assuming the fact that the SLE is a.s. a simple curve $\gamma_t = K_t \setminus K_{t-}$ for $\kappa < 4$). This is also to illustrate the type of techniques that is used to derive such properties of SLE:

Proposition 3.5 *For $\kappa < 4$, almost surely, $\lim_{t \to \infty} \gamma_t = \infty$.*

Loosely speaking, the SLE is transient. Actually (see [118]), this result is in fact valid for all κ, but the proof is (a little bit) more involved.

Proof. Let $\delta \in (0, 1/4)$, $x > 1$, and suppose that

$$t_\delta := \inf\{t > 0 \ : \ d(\gamma_t, [1, x]) \leq \delta\}$$

is finite. Let $z_\delta = \gamma_{t_\delta}$. Clearly, $g_{t_\delta}(z_\delta) = W_{t_\delta}$. Note that $g_{t_\delta}(1/2) - W_{t_\delta}$ is (up to a multiplicative constant) the limit when $y \to +\infty$ of y times the probability that a planar Brownian motion started from iy exits $\mathbb{H}$ in the interval $[W_{t_\delta}, g_{t_\delta}(1/2)]$. By conformal invariance, this is the same as the limit of y times the probability that a planar Brownian motion started from iy exits H_{t_δ} through the boundary of H_{t_δ} which is "between" z_δ and $1/2$. But in order to achieve this, the planar Brownian motion has in particular to hit the vertical segment joining z_δ to the real line before exiting $\mathbb{H}$. This segment has length at most δ. Hence,

$$|g_{t_\delta}(1/2) - W_{t_\delta}| \leq O(\delta).$$

On the other hand, $\lim_{t \to \infty}(g_t(1/2) - W_t) = \infty$ because $\kappa < 4$ (and the corresponding Bessel process is transient). It follows that a.s.,

$$d(\gamma[0, \infty], [1, x]) > 0.$$

By the scaling property and monotonicity, it follows that almost surely, for all $0 < x_1 < x_2$, the distance $d(\gamma[0, \infty], [x_1, x_2])$ is almost surely strictly positive.

  Let τ denote the hitting time of the unit circle by the SLE. Since $\mathbb{R} \cap \gamma[0, \tau] = \{0\}$, it follows that $0 \in \partial H_\tau$. For all $\varepsilon > 0$, there exists $0 < x_1 < x_2$ such that with probability at least $1 - \varepsilon$ the two images of 0 under g_τ are in $[W_\tau - x_2, W_\tau - x_1] \cup [W_\tau + x_1, W_\tau + x_2]$. It follows from the strong Markov property and from the previous result that with probability at least $1 - \varepsilon$,

$$d(g_\tau(\gamma[\tau, \infty)) - W_\tau, [-x_2, -x_1] \cup [x_1, x_2]) > 0.$$

Hence, it follows that in fact, almost surely

$$d(0, \gamma[\tau, \infty)) > 0$$

and the Lemma readily follows (for instance using the scaling property once again). $\square$

Bibliographical Comments

Again, many of the ideas in this chapter were contained or follow readily from Schramm's first paper [123]. Rohde-Schramm [118] have derived various almost sure properties of SLE (Hölder boundary, generated by a continuous path, transience). Proposition 3.3 is derived (in a more general setting) in [86]. It was Carleson who first noted that Cardy's formula (which Cardy predicted for crossing probabilities for critical percolation) has a simple interpretation in an equilateral triangle. The interpretation of the functions $\tilde{F}$ in terms of isocele triangles was pointed out by Dubédat [39]. Another justification to the fact that $\kappa = 6$ is the unique possible scaling limit of critical percolation exploration processes (for site percolation on the triangular lattice, or for bond percolation on the square lattice) uses the fact that for these models the probability of existence of a left-right crossing of a square must be $1/2$ (see [123]). For references on Bessel processes, stochastic calculus, see e.g. [59, 117].

4 Chordal SLE and Restriction

4.1 Image of SLE under Conformal Maps

Suppose now that $(K_t, t \geq 0)$ is chordal SLE_κ in the upper half-plane $\mathbb{H}$.

Definition. We say that a hull A that is at positive distance of the origin is a Hull (with capital H). When A is such a Hull, we define Φ_A the normalized conformal map from $\mathbb{H} \setminus A$ onto $\mathbb{H}$ as before. We also define Ψ_A the conformal map from $\mathbb{H} \setminus A$ onto $\mathbb{H}$ such that $\Psi(z) \sim z$ when $z \to \infty$ and $\Psi(0) = 0$. Note that $\Psi(z) = \Phi(z) - \Phi(0)$.

Let $A \subset \overline{\mathbb{H}}$ denote a Hull. Define $T = \inf\{t \; : \; K_t \cap A \neq \emptyset\}$ and for all $t < T$,

$$\tilde{K}_t := \Phi(K_t).$$

Let us immediately emphasize that the time-parametrization of K_t and therefore also of $\tilde{K}_t$ is given in terms of the "size" of $K_t = \Phi^{-1}(\tilde{K}_t)$ in $\mathbb{H}$ and not in terms of the "size" of $\tilde{K}_t$ itself in $\mathbb{H}$. One of the goals of this section is to study the evolution of $\tilde{K}_t$ and to compare it with that of K_t.

For $t < T$, we also define the conformal map h_t from $g_t(H_t \cap H)$ onto $\mathbb{H}$ (where $H = \mathbb{H} \setminus A$). Note that $h_0 = \Phi$. Since $g_t(A)$ is at positive distance of W_t for $t < T$, we can define

$$\tilde{W}_t = h_t(W_t).$$

Define finally also the normalized conformal map $\tilde{g}_t$ from $\Phi(H_t \cap H)$ onto $\mathbb{H}$. Note that (as long as $t < T$),

$$h_t \circ g_t = \tilde{g}_t \circ h_0.$$

In short, all these maps are normalized, $h_0 = \Phi$ removes A and $\tilde{g}_t$ removes $\tilde{K}_t$, while g_t removes K_t and h_t removes $g_t(A)$.

The family $(\tilde{K}_t, t < T)$ is a "continuously" growing family of subsets of $\mathbb{H}$ satisfying Proposition 2.1 except that a time-change is required in order to parametrize it as a Loewner chain. We therefore define the function

$$a(t) := a(A \cup K_t) = a(A) + a(\tilde{K}_t).$$

A simple time-change shows that

$$\partial \tilde{g}_t(z) = \frac{2\partial_t a}{\tilde{g}_t(z) - \tilde{W}_t}.$$

Hence, in order to understand the evolution of $\tilde{K}_t$, we have to understand the evolutions of $\tilde{W}_t$ and of $a(t)$.

The scaling rule $a(\lambda \cdot) = \sqrt{\lambda} a(\cdot)$ shows that

$$\partial_t a(t) = h_t'(W_t)^2.$$

On the other hand,

$$h_t = \tilde{g}_t \circ \Phi \circ g_t^{-1}$$

and

$$\partial_t(g_t^{-1}(z)) = -2\frac{(g_t^{-1})'(z)}{z - W_t}$$

so that putting the pieces together, we see that

$$\partial_t h_t(z) = \frac{2h_t'(W_t)^2}{h_t(z) - \tilde{W}_t} - \frac{2h_t'(z)}{z - W_t}. \tag{4.1}$$

Recall that $\tilde{W}_t = h_t(W_t)$. The previous formula is valid for all $z \in \mathbb{H} \setminus g_t(A)$. In fact, one can even extend it to $z = W_t$:

$$(\partial_t h_t)(W_t) = \lim_{z \to W_t} \left(\frac{2h_t'(W_t)^2}{h_t(z) - \tilde{W}_t} - \frac{2h_t'(z)}{z - W_t} \right) = -3h_t''(W_t)$$

(note that h_t is smooth near W_t because of Schwarz reflection). Itô's formula (this is not the classical formula since h_t is random, but it is adapted with respect to the filtration of W_t, it is C^1 with respect to t, so that Itô's formula still holds, see e.g., exercise IV.3.12 in [117]) can be applied:

$$d\tilde{W}_t = (\partial_t h_t)(W_t)dt + h_t'(W_t)dW_t + \frac{\kappa}{2}h_t''(W_t)dt.$$

Hence,

$$d\tilde{W}_t = h_t'(W_t)dW_t + [(\kappa/2) - 3]h_t''(W_t).$$

Clearly, the value $\kappa = 6$ will play a special role here. The next section is devoted to this case.

4.2 Locality for SLE_6

Throughout this section, we will assume that $\kappa = 6$. Then,

$$\tilde{W}_t = \int_0^t h_s'(W_s)dW_s.$$

Recall also that $a_t - a_0 = \int_0^t h_s'(W_s)^2 ds = \langle \tilde{W} \rangle_t$. Hence, if we define $(\hat{W}_a, a \geq 0)$ in such a way that

$$\tilde{W}_t = \hat{W}_{a(t)-a(0)},$$

then $\hat{W} - \hat{W}_0$ and W have the same law. If we define $\hat{g}_a$ in such a way that $\tilde{g}_t = \hat{g}_{a(t)}$, then

$$\partial_a \hat{g}_a(z) = \frac{2}{\hat{g}_a(z) - \hat{W}_a}.$$

Hence, modulo time-change, the evolution of $\tilde{K}_t - \hat{W}_0$ up to $t = T$ is that of chordal SLE_6. Suppose that $\tilde{T}$ is the first time at which K_t hits $\Phi(\partial A)$. We have just proved SLE_6's locality property:

Theorem 4.1. *Modulo time-reparametrization, the processes* $(\tilde{K}_t - \Phi(0), t < T)$ *and* $(K_t, t < \tilde{T})$ *have the same law.*

We now discuss some consequences of this result. Suppose first that

$$A = A_\varepsilon = \{e^{i\theta} \ : \ \theta \in [0, \pi - \varepsilon]\}.$$

Recall that $\Phi = \Phi_\varepsilon$ is the normalized map from $\mathbb{H} \setminus A$ onto $\mathbb{H}$. Let

$$\psi_\varepsilon(z) = \frac{\Phi_\varepsilon(z)}{\Phi'_\varepsilon(0)}.$$

It is easy to see that when $\varepsilon \to 0$, the mappings ψ_ε converge uniformly on any set $V_\delta := \{z \in \mathbb{H} \ : \ |z| < 1 - \delta\}$ towards the conformal map ψ from $V := \{z \in \mathbb{H} \ : \ |z| < 1\}$ onto $\mathbb{H}$ such that $\psi(0) = 0$, $\psi'(0) = 1$ and $\psi(-1) = \infty$. Theorem 4.1 shows that for each $\varepsilon > 0$, the law of the process $\psi_\varepsilon(K_t)$ up to its hitting time of $\psi_\varepsilon(A_\varepsilon)$ is a time-change of chordal SLE_6. In particular, letting $\varepsilon \to 0$ for each fixed $\delta > 0$ shows readily that:

Corollary 4.2. *Let* $(K_t, t \geq 0)$ *denote the law of chordal SLE_6 from 0 to -1 in V. Let T the first time at which K_t hits the unit circle. Then, the law of $(K_t, t < T)$ is identical (modulo time-change) to that of chordal SLE_6 in $\mathbb{H}$ (from 0 to ∞) up to its first hitting time of the unit circle.*

The same reasoning can be applied to $\{e^{i\theta} \ : \ \theta \in [\varepsilon, \pi]\}$ instead of A_ε. It shows that the law described in the corollary is also identical to that of chordal SLE_6 from 0 to $+1$ in V (up to the hitting time of the unit circle). By mapping the set V onto any other simply connected domain, we get the following splitting property:

Corollary 4.3. *Let* $D \subset \mathbb{H}$ *denote a simply connected subset of $\mathbb{H}$ such that the boundary of ∂D is a continuous Jordan curve. Let a, b, b' denote three distinct points on ∂D and call ∂ the connected component of $\partial D \setminus \{b, b'\}$ that does not contain a. Then: up to their first hitting times of ∂ and modulo time-change, the laws of chordal SLE_6 from a to b and from a to b' in D are identical.*

Note that these properties of chordal SLE_6 are not surprising if one thinks of SLE_6 as the scaling limit of critical percolation interfaces. They generalize the properties of hitting probabilities for SLE_6 that we derived in the previous chapter.

4.3 Restriction for $SLE_{8/3}$

We now apply the same technique as in the first subsection to understand how $h'_t(W_t)$ evolves. Recall that h_t is smooth in the neighbourhood of W_t by Schwarz reflection. Hence $h'_t(W_t)$ is a positive real (as long as $t < T$).

Differentiating Equation (4.1) with respect to z (this is licit as long as $t < T$) gives

$$\partial_t h_t'(z) = \frac{-2h_t'(W_t)^2 h_t'(z)}{(h_t(z) - \tilde{W}_t)^2} + \frac{2h_t'(z)}{(z - W_t)^2} - \frac{2h_t''(z)}{z - W_t}.$$

If we take the limit when $z \to W_t$, we get that

$$(\partial_t h_t')(W_t) = \frac{h_t''(W_t)^2}{2h_t'(W_t)} - \frac{4}{3}h_t'''(W_t).$$

Hence, Itô's formula (in its random version as before) shows that

$$d[h_t'(W_t)] = h_t''(W_t)dW_t + \left[\frac{h_t''(W_t)^2}{2h_t'(W_t)} + (\kappa/2 - 4/3)h_t'''(W_t) \right] dt.$$

This time, it is the value $\kappa = 8/3$ that plays a special role. Let us in this section from now on suppose that $\kappa = 8/3$. Then, we see that

$$d[h_t'(W_t)^{5/8}] = \frac{5h_t''(W_t)}{8h_t'(W_t)^{3/8}}dW_t.$$

The important feature is that the drift term disappear so that: $(h_t'(W_t)^{5/8}, t < T)$ is a local martingale. This has the following important consequence:

Proposition 4.4 *Consider chordal $SLE_{8/3}$ in $\mathbb{H}$. Then, for any Hull A,*

$$\mathbf{P}[\forall t \geq 0,\ K_t \cap A = \emptyset] = \Phi_A'(0)^{5/8}.$$

Proof. The quantity $M_t := h_t'(W_t)^{5/8}$ is a local martingale. Recall that h_t is a normalized map from a subset of $\mathbb{H}$ onto $\mathbb{H}$. Hence, for all $t < T$, $M_t \leq 1$ and M is a bounded martingale. We have to understand the behaviour of M_t when $t \to T$ in the two cases $T < \infty$ and $T = \infty$. When $T = \infty$, one can use the transience of the SLE: Define for each R, the hitting time τ_R of the circle of radius R. Then, simple considerations using harmonic measure for instance show that

$$\lim_{R \to \infty} h_{\tau_R}'(W_{\tau_R}) = 1.$$

In the case where $T < \infty$, one can for instance first approximate A by a Hull with a smooth boundary, and show that in this case, $\lim_{t \to T} h_t'(W_t) = 0$ for any path γ in the upper half-plane that hits A away from the real line. See [95] for details.

Finally, since M_t converges in L^1 and almost surely when $t \to T$, we get that $\mathbf{P}[T = \infty] = \mathbf{E}[M_T] = E[M_0] = \Phi'(0)^{5/8}$. $\qquad\square$

Let us now define the random set

$$K_\infty = \cup_{t > 0} K_t.$$

Corollary 4.5. *Suppose that A_0 is a Hull, then the conditional law of K_∞ given $K_\infty \cap A_0 = \emptyset$ is identical to the law of $\Psi_{A_0}^{-1}(K_\infty)$.*

Proof. Note that K_∞ is a closed set because of the transience of $(K_t, t \geq 0)$. The law of such a random set is characterized by the value of $\mathbf{P}[K_\infty \cap A = \emptyset]$ for all Hulls A (this set of events is a generating π-system of the σ-field on which we define K_∞). Suppose now that the Hull A_0 is fixed. By Proposition 4.4, K_∞ avoids A_0 with positive probability. Suppose that A is another Hull. Then

$$\mathbf{P}[\Psi_{A_0}(K_\infty) \cap A = \emptyset | K_\infty \cap A_0 = \emptyset]$$
$$= \frac{\mathbf{P}[K_\infty \cap (\mathbb{H} \setminus (\Psi_{A_0}^{-1} \circ \Psi_A^{-1}(\mathbb{H}))) = \emptyset]}{\mathbf{P}[K_\infty \cap A_0 = \emptyset]}$$
$$= \left(\frac{\Psi_{A_0}'(0)\Psi_A'(0)}{\Psi_{A_0}'(0)} \right)^{5/8}$$
$$= \mathbf{P}[K_\infty \cap A = \emptyset].$$

Since this is true for all Hull A, it follows that the the law of $\Psi_{A_0}(K_\infty)$ given $\{K_\infty \cap A_0 = \emptyset\}$ is identical to the law of K_∞. □

This striking property of $SLE_{8/3}$ has many nice consequences. It will enable us to relate it to the Brownian frontier in the next chapter. It also shows that it is the natural candidate for the scaling limit of planar self-avoiding walks. More precisely, one can show that when $n \to \infty$, the uniform measure on self-avoiding walks of length n in the upper half-plane $\mathbb{N} \times \mathbb{Z}$ started from the origin converges to a law of infinite self-avoiding walks. The conjecture is that the scaling limit of this infinite self-avoiding walk is $SLE_{8/3}$. See [94] for more on this. Note that there exist algorithms to simulate half-plane self-avoiding walks (see [60, 105]; Figure 4.1 is due to Tom Kennedy). The conjecture that the half-plane SAW scaling limit is chordal $SLE_{8/3}$ has recently been comforted by simulations [61].

Let us briefly conclude this chapter by mentioning the following characterization of $SLE_{8/3}$ that does not use explicitly Loewner's equation (even though its proof does):

Theorem 4.6. *Chordal $SLE_{8/3}$ is the unique measure on continuous simple curves γ from 0 to ∞ in $\mathbb{H}$ such that for all Hull, the law of γ conditioned to avoid A is identical to the law of $\Psi^{-1}(\gamma)$.*

The proof of this Theorem uses the complete description of all measures on simply connected closed sets (not necessarily curves) joining 0 to ∞ in $\mathbb{H}$ that satisfy this condition. These measures (called restriction measures in [95]) are constructed using SLE_κ (in fact, by adding Brownian bubbles to the SLE_κ paths) for other values of κ (in fact for $\kappa \in (0, 8/3]$) and it turns out that the only measure with these properties that is supported on simple curves is $SLE_{8/3}$.

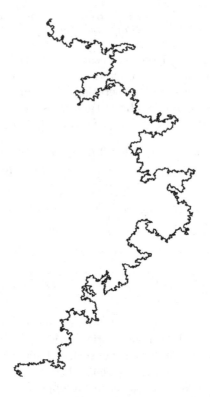

Fig. 4.1. Sample of the beginning of a half-plane walk (conjectured to converge to chordal $SLE_{8/3}$).

Bibliographical Comments

All the material of this chapter is borrowed from [95], to which we refer for further details. The locality property for SLE_6 was first proved in [87], using a different method. Restriction properties are closely related to conformal field theory [17, 18, 30, 31, 32, 34], as pointed out in [52, 53]. They have also interpretations in terms of highest-weight representations of the Lie algebra of polynomial vector fields on the unit circle. In fact, Theorem 4.6 corresponds to the fact that the unique such representation that is degenerate at level 2 has its highest weight equal to 5/8. See [52, 53].

5 SLE and the Brownian Frontier

5.1 A Reflected Brownian Motion

In this section, we introduce a two-dimensional Brownian motion with a certain oblique reflection on the boundary of a domain, and we will relate its outer boundary to that of SLE_6.

Let us first define this reflected Brownian motion in the upper half-plane $\mathbb{H}$. Define for any $x \in \mathbb{R}$, the vector $u(x) = \exp(i\pi/3)$ if $x \geq 0$ and $u(x) = \exp(2i\pi/3)$ if $x < 0$. It is the vector field with angle $2\pi/3$ pointing "away from the origin". Suppose that $Z_t^* = X_t^* + iY_t^*$ is an ordinary planar Brownian path started from 0. Then, there exists a unique pair (Z_t, ℓ_t) of continuous processes such that Z_t takes its values in $\overline{\mathbb{H}}$, ℓ_t is a non-decreasing real-valued continuous function with $\ell_0 = 0$ that increases only when $Z_t \in \mathbb{R}$, and

$$Z_t = Z_t^* + \int_0^t u(Z_s)d\ell_s.$$

The process $(Z_t, t \geq 0)$ is called the reflected Brownian motion in $\mathbb{H}$ with reflection vector field $u(\cdot)$. Note that the process Z in fact only depends on the direction of $u(\cdot)$ and not on its modulus. For instance Z is also the reflected Brownian motion in $\mathbb{H}$ with reflection vector field $2u(\cdot)$ (just change ℓ into $\ell/2$).

An equivalent way to define this process is to first define the reflected (one-dimensional) Brownian motion

$$Y_t = Y_t^* - \min_{s \in [0,t]} Y_s^*.$$

The local time at 0 of Y is simply $l_t = -\min_{[0,t]} Y^*$. Then, define X in such a way that

$$X_t = X_t^* + \int_0^t \text{sgn}(X_s)\frac{1}{\sqrt{3}}dl_s$$

and verify that $Z_t = X_t + iY_t$ satisfy the required conditions.

Brownian motion with oblique reflection on domains have been extensively studied, and this is not the proper place to review all results. We just mention that the general theory of such processes (e.g., [130]) ensures that the previously defined process Z^* exists.

Reflected planar Brownian motion (even with oblique reflection) are also invariant under conformal transformations. Suppose for instance that φ is a conformal transformation from a smooth subset V (such that $[-1, 1] \subset \partial V$) of $\mathbb{H}$ onto a smooth domain D. Recall that

$$Z_t = Z_t^* + \int_0^t u(Z_s)d\ell_s.$$

Define
$$\sigma_V := \inf\{t > 0 \; : \; \partial V \setminus (-1,1)\}.$$

Taylor-expanding each term in the sum

$$\varphi(Z_t) - \varphi(0) = \sum_{j=1}^{n}(\varphi(Z_{jt/n}) - \varphi(Z_{(j-1)t/n}))$$

just as in the proof of Itô's formula (letting $n \to \infty$), it follows (using the fact that the real and imaginary parts of φ are harmonic) that for all $t \le \sigma_V$,

$$\varphi(Z_t) = \int_0^t \varphi'(Z_s)dZ_s^* + + \int_0^t u(Z_s)\varphi'(Z_s)d\ell_s.$$

Hence, if one time-changes $\varphi(Z)$ using the clock $u(t) = \int_0^t |\varphi'(Z_s)|^2 ds$, we see that $\varphi(Z_u)$ is also a stopped reflected Brownian motion in D with the reflection vector field $(\varphi'(\varphi^{-1}(\cdot)) \times u(\varphi^{-1}(\cdot))$ on ∂D.

This has the following useful consequences: Suppose that $V \subset \mathbb{H}$ and σ_V are as before. Note that $\sigma_{\mathbb{H}}$ is the first time at which Z_t hits $\mathbb{R} \setminus (-1,1)$. There exists a unique conformal map φ from V onto $\mathbb{H}$ such that $\varphi(-1) = -1$, $\varphi(0) = 0$ and $\varphi(1) = 1$.

Lemma 5.1. *Modulo time-change, the laws of $(\varphi(Z_t), t \le \sigma_V)$ and of $(Z_t, t \le \sigma_{\mathbb{H}})$ are identical.*

In other words, The reflected Brownian motion Z satisfies the same locality property as SLE_6.

A slight modification of the above proof of conformal invariance for reflected Brownian motions shows that the image of Z under the conformal map $z \mapsto z^{1/3}$ from $\mathbb{H}$ onto the wedge

$$\mathcal{W} := \{re^{i\theta} \; : \; r > 0, \theta \in (0, \pi/3)\}$$

is reflected Brownian motion in that wedge, started from the origin, with reflection vector field $u(x) = e^{i\pi/3}$ on $\mathbb{R}_+$ and $u(x) = 1$ on $e^{i\pi/3}\mathbb{R}_+$. We use this observation to give a simple proof of the following fact on hitting probabilities for Z:

Lemma 5.2. *Suppose that Φ is the conformal transformation from $\mathbb{H}$ onto an equilateral triangle OAC such that $\Phi(0) = O$, $\Phi(-1) = A$ and $\Phi(1) = C$. Then, the law of $\Phi(Z_{\sigma_{\mathbb{H}}})$ is uniform on AC.*

Proof. One elementary convincing proof uses discrete approximations. Here is a brief outline of this proof: Define $\omega = \exp(i\pi/3)$. Consider a triangular grid in the wedge $\mathcal{W}$ i.e. $\{m + m'\omega \; : \; m, m' \ge 0\}$. Let $(S_n, n \ge 0)$ denote simple random walk on this grid that is started from 0. In the inside of $\mathcal{W}$, its transition probabilities are that of simple random walk (with probability 1/6 to jump to each of its neighbours). When S hits the (positive) real line at x, it has the following transition probabilities: $p(x, x+1) = 1/3$ and

$$p(x, x-1) = p(x, x+\omega) = p(x, x+\omega^2) = p(x, x) = \frac{1}{6}.$$

and the symmetric ones on $\omega\mathbb{N}$: $p(x, x+\omega) = 1/3$ and

$$p(x, x+1) = p(x, x+1/\omega) = p(x, x+1/\omega^2) = p(x, x) = \frac{1}{6}.$$

Finally, at the origin, $p(0, 1) = p(0, \omega) = 1/2$. It is not difficult to see that in the scaling limit, such a random walk converges to reflected Brownian motion in $\mathcal{W}$ with the reflection vector field $u(\cdot)$ on $\partial\mathcal{W}$. This is due to the fact that the bias of the simple random walk when it hits $\partial\mathcal{W}$ is proportional to u. Moreover, it is easy to check that if $S_0 = 0$, then if one writes $S_n = e^{i\pi/6}r_n + \omega^2 s_n$, then the conditional law of s_n given $(r_j, j \leq n)$ is the uniform distribution among the permitted values of s given r_n. In other words, the "uniform distribution of s is preserved, independently from r". In particular, the hitting distribution of the simple random walk S on the segment $N + \omega^2[0, N]$, is simply the uniform distribution on $\{N, N + \omega^2, N + 2\omega^2, \ldots, N + N\omega^2\}$. The Lemma follows, letting $N \to \infty$. $\qquad\square$

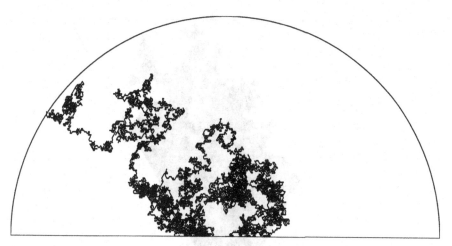

Fig. 5.1. The reflected Brownian motion stopped at its hitting time of the unit circle

We are now ready to state and prove the following result:

Theorem 5.3. *Define the following two sets:*

- *Consider chordal SLE_6 $(K_t, t \geq 0)$ in $\mathbb{H}$ (or in V) up to its first hitting time T of $\mathbb{R} \setminus (-1, 1)$. Let e denote the point at which the SLE hits $\mathbb{R} \setminus (-1, 1)$, and let $E := \{e\} \cup \cup_{t < T} K_t$.*

- *Consider the set of points F in $\overline{\mathbb{H}}$ that are disconnected (in $\mathbb{H}$) from $\mathbb{R} \setminus (-1, 1)$ by $Z[0, \sigma_{\mathbb{H}}]$.*

Then, the laws of E and of F are identical.

Proof. Note that Lemma 5.2, Lemma 5.1, Theorem 4.1 and Proposition 3.4 show that E and F both have the following properties:

- They are random compact sets that intersect $\mathbb{R} \setminus (-1, 1)$ at just one point x and the law of $\Phi(x)$ is uniform on AC.
- Their complement in $\overline{\mathbb{H}}$ consists of two connected components (one unbounded, one bounded).
- For all V as before, the probability that $E \subset V$ is identical to the probability that $\sigma_V = \sigma_{\mathbb{H}}$ (and the corresponding result for F).

If we combine these two properties, we see that for all such V,

$$\mathbf{P}[E \subset V] = \mathbf{P}[F \subset V] = \frac{\text{length}(\Phi \circ \varphi(\partial V \setminus \mathbb{R}))}{AC}$$

(this is because the law of the image under $\Phi \circ \varphi$ of the "hitting point" of $\partial V \setminus (-1, 1)$ is uniform on AC. But this determines completely the laws of E and of F and therefore implies that they are equal. $\qquad \square$

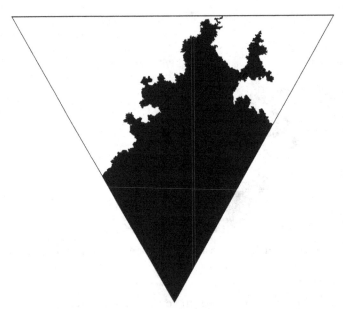

Fig. 5.2. The filling of RBM (or of the SLE_6 curve) in a triangle

Using conformal invariance, the previous result can be adapted in any domain. For instance, Figure 5.2 could represent both the filling of a reflected

Brownian motion (or of a SLE_6 curve), started at the bottom of the triangle stopped at their first hitting of the top segment. Recall that the law of this hitting point is uniformly distributed.

5.2 Brownian Excursions and $SLE_{8/3}$

We now describe a probability measure on Brownian excursions from 0 to infinity in $\mathbb{H}$ (which is closely related to the measures on excursions that were considered in [97]). One can view this measure on paths as the law of planar Brownian motion W (not to be confused with the $\sqrt{\kappa}B$ in the previous chapters) started from 0 and conditioned to stay in $\mathbb{H}$ at all positive times.

Let X and Y denote two independent processes such that X is standard one-dimensional Brownian motion and Y is a three-dimensional Bessel process (see e.g., [117] for background on three-dimensional Bessel processes, its relation to Brownian motion conditioned to stay positive and stochastic differential equations) that are both started from 0. Let us briefly recall that a three-dimensional Bessel process is the modulus of a three-dimensional Brownian motion, and that it can be defined as the solution to the stochastic differential equation

$$dY_t = dw_t + \frac{1}{Y_t}dt$$

(where w is one-dimensional standard Brownian motion). It is very easy to see that $(1/Y, t \geq t_0)$ is a local martingale for all $t_0 > 0$, and that if T_r denotes the hitting time of r by Y, then the law of $(Y_{T_r+t}, t < T_R - T_r)$ is identical to that of a Brownian motion started from r and conditioned to hit R before 0 (if $0 < r < R$). Loosely speaking Y is a Brownian motion started from 0 and conditioned to stay forever positive. Note that almost surely $\lim_{t\to\infty} Y_t = \infty$.

We now define $W = X + iY$. In other words, W has the same law as the solution to the following stochastic differential equation:

$$dW_t = d\beta_t + i\frac{1}{\Im(W_t)}dt \tag{5.1}$$

with $W_0 = 0$, where β is a complex-valued Brownian motion. Note that W is a strong Markov process. Let T_r denote the hitting time of the line $\mathbb{R} + ir$ by this process W (i.e., the hitting time of r by X). Let S denote a random variable with the same law as W_{T_1}. Then, scaling and the relation between one-dimensional Brownian motion conditioned to stay positive and the three-dimensional Bessel process shows immediately that for all $0 < r < R$, the law of $W[T_r, T_R]$ is the law of a Brownian motion started with the same law as rS, stopped at its first hitting of $iR + \mathbb{R}$, and conditioned to stay in the upper half-plane up to that time. Note that the probability of this event is r/R.

By mapping conformally $\mathbb{H}$ onto any other simply connected domain D ($D \neq \mathbb{C}$), and looking at the image of the Brownian excursion in $\mathbb{H}$ under this map, one gets the law of a Brownian excursion in D from the image of 0 to

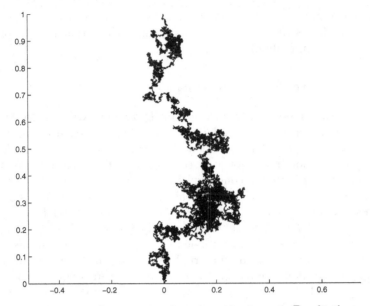

Fig. 5.3. An excursion from 0 to i in the strip $\mathbb{R} \times [0, 1]$

the image of ∞. As for SLE, this law is well-defined up to linear time-change. One can also directly define this excursion in D as the solution to a stochastic differential equation "forcing the Brownian motion to hit ∂D at the image of infinity."

The following result was observed by Bálint Virág [129] (see also [97, 95]):

Lemma 5.4. *Suppose A is a Hull and W is a Brownian excursion in $\mathbb{H}$ from 0 to ∞. Then $\mathbf{P}[W[0, \infty) \cap A = \emptyset] = \Phi'_A(0)$.*

Proof. Suppose that W is a solution to (5.1) started from $z \in \Phi^{-1}(\mathbb{H})$. Let Z denote a planar Brownian motion started from z. Let $\tau_R(V)$ denote the hitting time of $iR + \mathbb{R}$ by a process V. When $\Im(z) \to \infty$, $\Im(\Phi(z)) = \Im(z) + o(1)$, and it therefore follows easily from the strong Markov property of planar Brownian motion that when $R \to \infty$,

$$\mathbf{P}[\Phi(Z)[0, \tau_R(Z)] \subset \mathbb{H}] \sim \mathbf{P}[\Phi(Z)[0, \tau_R(\Phi(Z))]] \subset \mathbb{H}].$$

But since $\Phi(Z)$ is a time-changed Brownian motion, the right-hand probability is equal to $\Im(\Phi(z))/R$, so that

$$\mathbf{P}[W[0, \tau_R(W)] \subset \Phi^{-1}(\mathbb{H})] = \frac{\mathbf{P}[Z[0, \tau_R(Z)] \subset \Phi^{-1}(\mathbb{H})]}{\mathbf{P}[Z[0, \tau_R(Z)] \subset \mathbb{H}]} = \frac{\Im\Phi(z)}{\Im(z)} + o(1)$$

when $R \to \infty$. In the limit $R \to \infty$, we get

$$\mathbf{P}[W[0, \infty) \subset \Phi^{-1}(\mathbb{H}) \mid W_0 = z] = \frac{\Im\Phi(z)}{\Im(z)}. \tag{5.2}$$

When $z \to 0$, $\Phi(z) = z\Phi'(0) + O(|z|^2)$ so that

$$\begin{aligned}
\mathbf{P}[W[0,\infty) \subset \Phi^{-1}(\mathbb{H})] &= \lim_{r \to 0} \mathbf{P}[W[T_r,\infty) \subset \Phi^{-1}(\mathbb{H})] \\
&= \lim_{r \to 0} \mathbf{E}[\Im(\Phi(rA))/\Im(rA)] \\
&= \Phi'(0)
\end{aligned}$$

(one can use dominated convergence here since $\Im(\Phi(z)) \leq \Im(z)$ for all z). $\quad\square$

We now define the filling $\mathcal{H}$ of $W[0,\infty)$ as the set of points in $\overline{\mathbb{H}}$ that are disconnected from $\mathbb{R}$ by $W[0,\infty)$. This set is obtained by filling in all the bounded connected components of the complement of the curve W. Then, $\mathcal{H}$ is a closed unbounded set and $\mathbb{H} \setminus \mathcal{H}$ consists of two open connected components (with $[0,\infty)$ and $(-\infty,0]$ on their respective boundaries). The law of such a random set is characterized by the values of $\mathbf{P}[\mathcal{H} \cap A = \emptyset]$, where A spans all Hulls, because this family of events turn out to generate the σ-field on which $\mathcal{H}$ is defined, and to be stable under finite intersections. Hence, as in the case of K_∞ for $SLE_{8/3}$, the fact that

$$\mathbf{P}[\mathcal{H} \cap A = \emptyset] = \Phi'(0) \tag{5.3}$$

characterizes the law of $\mathcal{H}$ and yields that $\mathcal{H}$ also satisfies Corollary 4.5.

Theorem 5.5. *Suppose that $\mathcal{H}_8$ denotes the filling of the union of 8 independent chordal $SLE_{8/3}$'s. Suppose that $\mathcal{H}_5$ denotes the filling of the union of 5 independent Brownian excursions. Then, $\mathcal{H}_5$ and $\mathcal{H}_8$ have the same law.*

Proof. This is simply due to the fact that for all Hull A

$$\mathbf{P}[\mathcal{H}_5 \cap A = \emptyset] = \mathbf{P}[\mathcal{H}_8 \cap A = \emptyset] = \Phi'_A(0)^5$$

and that this characterizes these laws. $\quad\square$

This has various nice consequences (see [95]), some of which we now heuristically describe: First, since the boundary of $\mathcal{H}_8$ consists of the union of some parts of the $SLE_{8/3}$ curves, it follows that "locally", the outer boundary of a Brownian excursion (and therefore also of a Brownian motion) looks like one $SLE_{8/3}$ path. In the previous section, we did see that the outer boundaries of reflected Brownian motion and of SLE_6 are the same. Hence, "locally", the outer frontiers of SLE_6 and of planar Brownian motion look like an $SLE_{8/3}$ curve. Furthermore, since $SLE_{8/3}$ is symmetric, this shows that one cannot distinguish the inside from the outside of a planar Brownian curve by only seeing a part of its frontier. Since $SLE_{8/3}$ is conjectured to be the scaling limit of self-avoiding walks, this would also show that the Brownian frontier looks locally like the scaling limit of long self-avoiding curves (see [94]).

Bibliographical Comments

The idea that conformal invariance and restriction defines measures on random sets and makes it possible to understand the Brownian frontier in terms of other models (or the corresponding exponents) first appears in [97]. Most of the material of this chapter is borrowed from [95].

A discussion of the conjectured relation between $SLE_{8/3}$ and planar self-avoiding walks is discussed in [94]; one can in particular recover the predictions of Nienhuis [110] on the critical exponents for self-avoiding walks using SLE arguments.

The fact that the Brownian frontier had the same dimension as the scaling limit of self-avoiding walks was first observed visually by Mandelbrot [107].

6 Radial SLE

6.1 Definitions

Motivated by the example of LERW (among others) given in the introductory chapter, we now want to find a nice way to encode growing families of compact subsets $(K_t, t \geq 0)$ of the closed unit disk that are growing from the boundary point 1 towards 0. As in the chordal case, we are in fact going to focus on the conformal geometry of the complement H_t of K_t in the unit disc $\mathbb{U}$. One first has to find a natural time-parametrization. It turns out to be convenient to define the conformal map g_t from H_t onto $\mathbb{U}$ that is normalised by

$$g_t(0) = 0 \text{ and } g'_t(0) > 0.$$

Note that $g'_t(0) \geq 1$. This can be for instance derived using the fact that $\log g'_t(0)$ is the limit when $\varepsilon \to 0$ of $\log(1/\varepsilon)$ times the probability that a planar Brownian motion started from ε hits the circle of radius ε^2 before exiting H_t (an analyst would find this justification very strange, for sure).

Then (and this is simply because with obvious notation, $(\tilde{g}_s \circ g_t)(0) = \tilde{g}_s(0) \circ g'_t(0)$), one measures the "size" $a(K_t)$ of K_t via the derivative of g_t at the origin:

$$g'_t(0) = \exp(a(t)).$$

Hence, we will consider growing families of compact sets such that $a(K_t) = t$.

Suppose now that $(\zeta_t, t \geq 0)$ is a continuous function on the unit circle $\partial \mathbb{U}$. Define for all $z \in \bar{\mathbb{U}}$, the solution $g_t(z)$ to the ODE

$$\partial_t g_t(z) = -g_t(z) \frac{g_t(z) + \zeta_t}{g_t(z) - \zeta_t} \tag{6.1}$$

such that $g_0(z) = z$. This solution is well-defined up to the (possibly infinite) time $T(z)$ defined by

$$T(z) = \sup\{t > 0 : \min_{s \in [0,t)} |g_s(z) - \zeta_s| > 0\}.$$

We then define

$$K_t := \{z \in \bar{\mathbb{U}} : T(z) \leq t\}$$

and

$$U_t := \mathbb{U} \setminus K_t.$$

The family $(K_t, t \geq 0)$ is called the (radial) Loewner chain associated to the driving function ζ.

The general statements that we described in the chordal case are also valid in this radial case. One can add one feature that has no analog in the chordal case: It is possible to estimate the Euclidean distance d_t from 0 to K_t in terms of $a(t) = t$. Indeed, since U_t contains the disc $d_t \times \mathbb{U}$, it is clear that $g'_t(0) \leq 1/d_t$. On the other hand, a classical result of the theory of conformal mappings known as Koebe's 1/4 Theorem states that (if $a(K_t) = t$) $1/d_t \leq 4g_t'(0)$. This is loosely speaking due to the fact that the best K_t can do to get as close to 0 in "time t" is to shoot straight i.e. to choose $\zeta = 1$. Hence, for all $t \geq 0$,

$$e^{-t}/4 \leq d(0, K_t) \leq e^{-t}. \tag{6.2}$$

This will be quite useful later on.

Radial SLE_κ is then simply the random family of sets $(K_t, t \geq 0)$ that is obtained when

$$\zeta_t = \exp(i\sqrt{\kappa} B_t)$$

where $\kappa > 0$ is fixed and $(B_t, t \geq 0)$ is standard one-dimensional Brownian motion.

As in the chordal case, one can then define radial SLE from $a \in \partial D$ to $b \in D$ in any open simply connected domain D by taking the image of radial SLE in $\mathbb{U}$ under the conformal map Φ from $\mathbb{U}$ onto D such that $\Phi(1) = a$ and $\Phi(0) = b$. Note that this time, the time-parametrization is also well-defined since there exists only one such conformal map (recall that in the chordal case, one had to invoke the scaling property to make sure that chordal SLE in other domains than the half-space was properly defined).

6.2 Relation between Radial and Chordal SLE

In this section, we show that chordal SLE and radial SLE are very closely related. Let us start with the special case $\kappa = 6$.

Theorem 6.1. *Suppose that $x \in (0, 2\pi)$. Let $(K_t, t \geq 0)$ be a radial SLE_6 process. Set*

$$T := \inf\{t \geq 0 \ : \ \exp(ix) \in K_t\}.$$

Let $(\tilde{K}_u, u \geq 0)$ be a chordal SLE_6 process in $\mathbb{U}$ starting also at 1 and growing towards $\exp(ix)$, and let

$$\tilde{T} := \inf\{u \geq 0 \ : \ 0 \in \tilde{K}_u\}.$$

Then, up to a random time change, the process $t \mapsto K_t$ restricted to $[0, T)$ has the same law as the process $u \mapsto \tilde{K}_u$ restricted to $[0, \tilde{T})$.

Note that T (resp. $\tilde{T}$) is the first time where K_t (resp. $\tilde{K}_u$) disconnects 0 from 1.

When $\kappa \neq 6$, a weaker form of equivalence holds:

Proposition 6.2 *Let $(K_t, t \geq 0)$, $(\tilde{K}_u, u \geq 0)$, T and $\tilde{T}$ be defined just as in Theorem 6.1, except that they are SLE with general $\kappa > 0$. There exist two nondecreasing families of stopping times $(T_n, n \geq 1)$ and $(\tilde{T}_n, n \geq 1)$ such that almost surely, $T_n \to T$ and $\tilde{T}_n \to \tilde{T}$ when $n \to \infty$, and such that for each $n \geq 1$, the laws of $(K_t, t \in [0, T_n])$ and $(\tilde{K}_u, u \in [0, \tilde{T}_n])$ are equivalent (in the sense that they have a positive density with respect to each other) modulo increasing time change.*

These results imply that the properties of chordal SLE such as "being generated by a continuous curve" are also valid for radial SLE.

We prove both results simultaneously:

Proof. Let us first briefly recall how $\tilde{K}_u$ is defined. For convenience, we will restrict ourselves to $x = \pi$ (the proof in the general case is almost identical). Define the conformal map

$$\psi(z) = i\frac{1-z}{1+z}$$

from $\mathbb{U}$ onto $\mathbb{H}$ that satisfies $\psi(-1) = \infty$, $\psi(1) = 0$, and $\psi(0) = i$. Suppose that $u \mapsto \tilde{B}_u$ is a real-valued Brownian motion such that $\tilde{B}_0 = 0$. For all $z \in \mathbb{U}$, define the function $\tilde{g}_u = \tilde{g}_u(z)$ such that $\tilde{g}_0(z) = \psi(z)$ and

$$\partial_u \tilde{g}_u = \frac{2}{\tilde{g}_u - \sqrt{\kappa}\tilde{B}_u}.$$

This function is defined up to the (possibly infinite) time $\tilde{T}_z$ where $\tilde{g}_u(z)$ hits $\sqrt{\kappa}\tilde{B}_u$. Then, $\tilde{K}_u$ is defined by $\tilde{K}_u = \{z \in \mathbb{U} : \tilde{T}_z \leq u\}$, so that $\tilde{g}_u$ is a conformal map from $\mathbb{U} \setminus \tilde{K}_u$ onto the upper half-plane. This defines the process $(\tilde{K}_u, u \geq 0)$.

We are now going to compare it to radial SLE. Let $g_t : \mathbb{U} \setminus K_t \to \mathbb{U}$ be the conformal map normalized by $g_t(0) = 0$ and $g_t'(0) > 0$. Recall that

$$\partial_t g_t(z) = g_t(z)\frac{\zeta_t + g_t(z)}{\zeta_t - g_t(z)}, \tag{6.3}$$

where $\zeta_t = \exp(i\sqrt{\kappa}B_t)$, and B is Brownian motion on $\mathbb{R}$ with $B_0 = 0$. Let ψ be the same conformal map as before, and define

$$e_t := g_t(-1),$$
$$f_t(z) := \psi\big(g_t(z)/e_t\big),$$
$$r_t := \psi\big(\zeta_t/e_t\big).$$

These are well defined, as long as $t < T$. Note that f_t is a conformal map from $\mathbb{U} \setminus K_t$ onto the upper half-plane, $f_t(1) = \infty$, and $r_t \in \mathbb{R}$. From (6.3) it follows that

$$\partial_t f = -\frac{(1+r^2)(1+f^2)}{2(r-f)}.$$

Let

$$\varphi_t(z) = a(t)z + b(t)$$

where

$$a(0) = 1, \qquad \partial_t a = -(1+r^2)a/2$$

and

$$b(0) = 0, \qquad \partial_t b = -(1+r^2)ar/2.$$

Set

$$h_t := \varphi_t \circ f_t ,$$
$$\beta_t := \varphi_t\big(r(t)\big) .$$

Then (and this is the reason for the choice of the functions a and b)

$$\partial_t h = -(a/2)\frac{(1+r^2)^2}{r-f} = -\frac{(1+r^2)^2 a^2/2}{\beta - h} .$$

h_t is also a conformal map from $\mathbb{U} \setminus K_t$ onto the upper half-plane with $h_t(1) = \infty$. Note also that $h_0(z) = \psi(z)$. We introduce a new time parameter $u = u(t)$ by setting

$$\partial_t u = (1+r^2)^2 a^2/4, \qquad u(0) = 0 .$$

Then

$$\frac{\partial h}{\partial u} = \frac{-2}{\beta - h} .$$

Since this is the equation defining the chordal SLE process, it remains to show that $u \mapsto \beta_{t(u)}/\sqrt{\kappa}$ is related to Brownian motion (stopped at some random time). This is a direct but tedious application of Itô's formula:

$$dr_t = \frac{(1+r^2)\sqrt{\kappa}}{2} dB_t + \frac{r(1+r^2)}{2}\left(\frac{\kappa}{2} - 1\right) dt$$

and

$$d\beta_t = \frac{(1+r^2)a}{2}\left(\sqrt{\kappa}\, dB_t + (-3 + \frac{\kappa}{2})r\, dt\right) .$$

When $\kappa = 6$, the drift term disappears and this proves Theorem 6.1. When $\kappa \neq 6$, the drift term does not disappear. However, the law of $u \mapsto \beta_{t(u)}$ is absolutely continuous with respect to that of $\sqrt{\kappa}$ times a Brownian motion, as long as r and u remain bounded. More precisely: It suffices to take

$$T_n = \min\Big\{n, \inf\{t > 0 : |\zeta_t - e_t| < 1/n\}\Big\} .$$

Before T_n, $|r|$ remains bounded, a is bounded away from 0 (note also that $a \leq 1$ always), so that t/u is bounded and bounded away from 0. Hence, $u(T_n)$ is also bounded (since $T_n \leq n$).

It now follows directly from Girsanov's Theorem (see e.g., [117]) that the law of $\big(\beta(u)/\sqrt{\kappa}\big)_{u \leq u(T_n)}$ is equivalent to that of Brownian motion up to some (bounded) stopping time, and Proposition 6.2 follows. □

6.3 Radial SLE_6 and Reflected Brownian Motion

If one combines the radial-chordal equivalence for SLE_6 with the locality property for chordal SLE_6, one gets immediately a locality property for radial SLE_6, and the relation between fillings of radial SLE_6 and of reflected

Brownian motion. We do not state the locality property here (and leave it to the interested reader), but we state the relation between fillings of radial SLE and of reflected Brownian motions that we will use in the next chapters.

Before that, we have to say some words about how this reflected Brownian motion is defined in the unit disc. Suppose that $(Z_t, t \geq 0)$ is the reflected Brownian motion in the upper half-plane with reflection angle $2\pi/3$ away from the origin as in the previous chapter. Let us now define

$$\tilde{Z}_t := \exp(-iZ_t)$$

so that $\tilde{Z}$ takes its values in the unit disk and is started from $\tilde{Z}_0 = 1$. Clearly, since $\tilde{Z}_t \neq 0$ for all t, one can define the continuous version of its argument $(\theta_t, t \geq 0)$. Conformal invariance of planar Brownian motion shows that $\tilde{Z}_t$ behaves like (time-changed) Brownian motion as long as it stays away from the unit circle, and when it hits the unit circle, then it is reflected with angle $2\pi/3$ in the direction that "increases" $|\theta|$. Define

$$\tilde{\sigma}(r) := \inf\{t > 0 \ : \ |\tilde{Z}_t| = r\}$$

which is also the first time at which the imaginary part of Z hits $\log(1/r)$.

Theorem 6.3. *Suppose that $r < 1$. Define the two following random hulls:*

- *Suppose that $(K_t, t \geq 0)$ is radial SLE_6 as before. Let τ_r denote the first time at which radial K_t intersects the circle $\{|z| = r\}$. Define the event $\mathcal{H}(x, \tau_r)$ that K_{τ_r} does not disconnect 0 from $\exp(ix)$.*
- *On the event $\tilde{\mathcal{H}}(x, \tilde{\sigma}_r)$ that $\tilde{Z}[0, \tilde{\sigma}_r]$ does not disconnect 0 from $\exp(ix)$, define the connected component H of $\mathbb{U} \setminus \tilde{Z}$ that contains 0, and the hull $\tilde{K}_{\sigma_r} = \overline{\mathbb{U} \setminus H}$.*

Then, the two random sets $1_{\mathcal{H}(x,\tau_r)} K_{\tau_r}$ and $1_{\tilde{\mathcal{H}}_r(x,\tilde{\sigma}_r)} \tilde{K}_{\tilde{\sigma}_r}$ have the same law.

In particular,

$$\mathbf{P}[\mathcal{H}(x, \tau_r)] = \mathbf{P}[\tilde{\mathcal{H}}(x, \tilde{\sigma}_r))].$$

This shows that one can compute non-disconnection probabilities for reflecting Brownian motions using radial SLE_6.

Bibliographical Comments

For basic results on Loewner's equation, and basic complex analysis, we refer again to [1, 2, 49, 58]. The radial-chordal equivalence for SLE_6 has been derived in [87].

7 Some Critical Exponents for SLE

7.1 Disconnection Exponents

In this section, we fix $\kappa > 4$, and we consider radial SLE_κ in the unit disc started from 1. Our goal will be to estimate probabilities of events like

$$\mathcal{H}(x,t) = \{\exp(ix) \in \partial H_t\}$$

that K_t has not swallowed the point $\exp(ix) \in \partial \mathbb{U}$ from 0 at time t. Let us define the numbers

$$q_0 = q_0(\kappa) := 1 - \frac{4}{\kappa}$$

and

$$\lambda_0 = \lambda_0(\kappa) := \frac{\kappa}{8} - \frac{1}{2}.$$

Proposition 7.1 *There exists a constant c such that for all $t \geq 1$ and for all $x \in (0, 2\pi)$,*

$$e^{-\lambda_0 t}(\sin(x/2))^{q_0} \leq \mathbf{P}[\mathcal{H}(x,t)] \leq ce^{-\lambda_0 t}(\sin(x/2))^{q_0}.$$

Proof. We will use the notation

$$f(x,t) = \mathbf{P}[\mathcal{H}(x,t)].$$

Let $\zeta_t = \exp(i\sqrt{\kappa}B_t)$ be the driving process of the radial SLE_κ, with $B_0 = 0$. For all $x \in (0, 2\pi)$, let Y_t^x be the continuous real-valued function of t which satisfies

$$g_t(e^{ix}) = \zeta_t \exp(iY_t^x)$$

and $Y_0^x = x$. The function Y_t^x is defined on the set of pairs (x,t) such that $\mathcal{H}(x,t)$ holds. Since g_t satisfies Loewner's differential equation

$$\partial_t g_t(z) = g_t(z)\frac{\zeta_t + g_t(z)}{\zeta_t - g_t(z)}, \tag{7.1}$$

we find that

$$dY_t^x = \sqrt{\kappa}\, dB_t + \cot(Y_t^x/2)\, dt. \tag{7.2}$$

Let

$$\tau^x := \inf\left\{t \geq 0 : Y_t^x \in \{0, 2\pi\}\right\}$$

denote the time at which $\exp(ix)$ is absorbed by K_t, so that

$$f(x,t) = \mathbf{P}[\tau_x > t].$$

We therefore want to estimate the probability that the diffusion Y^x (started from x) has not hit $\{0, 2\pi\}$ before time t as $t \to \infty$. This is a standard problem.

The general theory of diffusion processes can be used to argue that $f(x,t)$ is smooth on $(0, 2\pi) \times \mathbb{R}_+$, and Itô's formula immediately shows that

$$\frac{\kappa}{2}\partial_x^2 f + \cot(x/2)\partial_x f = \partial_t f. \tag{7.3}$$

Moreover, for instance comparing Y with Bessel processes when Y is small, one can easily see that (here we use that $\kappa > 4$) for all $t > 0$,

$$\lim_{x \to 0+} f(x,t) = \lim_{x \to 2\pi-} f(x,t) = 0. \tag{7.4}$$

Hence, f is solution to (7.3) with boundary values (7.4) and $f(x,0) = 1$. This in fact characterizes f, and its long-time behaviour is described in terms of the first eigenvalue of the operator $\kappa\partial_x^2/2 + \cot(x/2)\partial_x$. More precisely, define

$$F(x,t) = \mathbf{E}[1_{\mathcal{H}(x,t)} \sin(Y_t^x/2)^{q_0}].$$

Then, it is easy to see that F also solves (7.3) with boundary values (7.4) but this time with initial data $F(x,0) = \sin(x/2)^{q_0}$. One can for instance invoke the maximum principle to construct a handcraft proof (as in [86]) of the fact that this characterizes F. Since $e^{-\lambda_0 t} \sin(x/2)^{q_0}$ also satisfies these conditions, it follows that

$$F(x,t) = e^{-\lambda_0 t} \sin(x/2)^{q_0}.$$

Hence,

$$f(x,t) = \mathbf{P}[\mathcal{H}(x,t)] \geq \mathbf{E}[1_{\mathcal{H}(x,t)} \sin(Y_t^x/2)^{q_0}] = e^{-\lambda_0 t} \sin(x/2)^{q_0}.$$

To prove the other inequality, one can for instance use an argument based on Harnack-type considerations: For instance, one can see that (uniformly in x) a positive fraction of the paths $(Y_t^x, t \in [0,1])$ such that $\tau_x > 1$ satisfy $Y_1^x \in [\pi/2, 3\pi/2]$. This then implies readily (using the Markov property at time $t - 1$) that for all $t \geq 1$,

$$f(x,t) \leq c_0\mathbf{P}[\tau_x > t \text{ and } Y_t^x \in [\pi/2, 3\pi/2]] \leq c_1 F(x,t) = c_1 e^{-\lambda_0 t} \sin(x/2)^{q_0}.$$

$\square$

7.2 Derivative Exponents

The previous argument can be generalized in order to derive the value of other exponents that will be very useful later on: We will focus on the moments of the derivative of g_t at $\exp(ix)$ on the event $\mathcal{H}(x,t)$. Note that on a heuristic level, $|g_t'(e^{ix})|$ measures how "far" e^{ix} is from the origin in H_t.

More precisely, we fix $b \geq 0$, and we define

$$f(x,t) := \mathbf{E}\left[|g_t'(\exp(ix))|^b 1_{\mathcal{H}(x,t)}\right].$$

We also define the numbers

$$q = q(\kappa, b) := \frac{\kappa - 4 + \sqrt{(\kappa - 4)^2 + 16b\kappa}}{2\kappa}$$

$$\lambda = \lambda(\kappa, b) := \frac{8b + \kappa - 4 + \sqrt{(\kappa - 4)^2 + 16b\kappa}}{16}.$$

The main result of this Section is the following generalization of Proposition 7.1:

Proposition 7.2 *There is a constant $c > 0$ such that for all $t \geq 1$, for all $x \in (0, 2\pi)$,*

$$e^{-\lambda t}\left(\sin(x/2)\right)^q \leq f(x, t) \leq ce^{-\lambda t}\left(\sin(x/2)\right)^q$$

Proof. We can assume that $b > 0$ since the case $b = 0$ was treated in the previous section. Let Y_t^x be as before and define for all $t < \tau^x$

$$\Phi_t^x := \left|g_t'(\exp(ix))\right|.$$

On $t \geq \tau^x$ set $\Phi_t^x := 0$. Note that on $t < \tau^x$

$$\Phi_t^x = \partial_x Y_t^x.$$

By differentiating (7.1) with respect to z, we find that for $t < \tau^x$

$$\partial_t \log \Phi_t^x = -\frac{1}{2\sin^2(Y_t^x/2)} \tag{7.5}$$

and hence (since $\Phi_0^x = 1$),

$$(\Phi_t^x)^b = \exp\left(-\frac{b}{2}\int_0^t \frac{ds}{\sin^2(Y_s^x/2)}\right), \tag{7.6}$$

for $t < \tau^x$. So, we can rewrite

$$f(x, t) = \mathbf{E}\left[1_{\mathcal{H}(x,t)} \exp\left(-\frac{b}{2}\int_0^t \frac{ds}{\sin^2(Y_s^x/2)}\right)\right].$$

Again, it is not difficult to see that the right hand side of (7.6) is 0 when $t = \tau^x$ and that

$$\lim_{x \to 0} f(x, t) = \lim_{x \to 2\pi} f(x, t) = 0 \tag{7.7}$$

holds for all fixed $t > 0$.

Let $F : [0, 2\pi] \to \mathbb{R}$ be a continuous function with $F(0) = F(2\pi) = 0$, which is smooth in $(0, 2\pi)$, and set

$$h(x, t) = h_F(x, t) := \mathbf{E}\left[(\Phi_t^x)^b F(Y_t^x)\right].$$

By (7.6) and the general theory of diffusion Markov processes, we know that h is smooth in $(0, 2\pi) \times \mathbb{R}_+$. From the Markov property for Y_t^x and (7.6), it follows that $h(Y_t^x, t' - t)(\Phi_t^x)^b$ is a local martingale on $t < \min\{\tau^x, t'\}$. Consequently, the drift term of the stochastic differential $d\big(h(Y_t^x, t'-t)(\Phi_t^x)^b\big)$ is zero at $t = 0$. By Itô's formula, this means

$$\partial_t h = \Lambda h, \tag{7.8}$$

where

$$\Lambda h := \frac{\kappa}{2} \partial_x^2 h + \cot(x/2)\, \partial_x h - \frac{b}{2\sin^2(x/2)}\, h\,.$$

We therefore choose

$$F(x) := \big(\sin(x/2)\big)^q,$$

and note that $F(x)e^{-\lambda t} = h_F$ because both satisfy (7.8) on $(0, 2\pi) \times [0, \infty)$, and have the same boundary values. Finally, one can conclude using the same type of argument as in Proposition 7.1. $\square$

7.3 First Consequences

Recall that for all $t \geq 0$, $d(0, K_t)e^t \in [1/4, 1]$. Hence, if τ_r denotes the hitting time of the circle of radius $r < 1$ by the radial SLE_κ, then $re^{\tau_r} \in [1/4, 1]$. Combining this with Propositions 7.1 and 7.2 then implies that for all fixed $\kappa > 4$, all $b \geq 0$, if λ, q are defined as before, there exists two positive finite constants c_1 and c_2 such that for all $r < r_0$,

$$c_1 r^\lambda (\sin(x/2))^q \leq \mathbf{E}\left[1_{\mathcal{H}(x,\tau_r)}|g'_{\tau_r}(\exp(ix))|^b\right] \leq c_2 r^\lambda (\sin(x/2))^q \tag{7.9}$$

(we used also the fact that $|g'_t(\exp(ix))|$ is an decreasing function of t).

When $b = 1$, one can note that

$$l_t := \int_0^{2\pi} dx |g'_t(e^{ix})| 1_{\mathcal{H}(x,t)}$$

is simply the length of the image under g_t of the arc $A_t := \partial H_t \cap \partial \mathbb{U}$ on the unit circle that have not yet been swallowed by K_t. In particular, if one starts a planar Brownian motion from 0, it has a probability $l_t/2\pi$ to hit the unit circle on the arc $g_t(A_t)$. By conformal invariance of planar Brownian motion, we see that $l_t/2\pi$ is also the probability that a planar Brownian motion started from 0 hits the unit circle before hitting K_t. Let Z denote planar Brownian motion, stopped at its hitting time σ of the unit circle. Integrating Proposition 7.2 for $x \in [0, 2\pi]$ therefore shows that there exist constants c'_1 and c'_2 such that (if K_t is radial SLE_6)

$$c'_1 r^{5/4} \leq \mathbf{P}[Z[0, \sigma] \cap K_{\tau_r} = \emptyset] \leq c'_2 r^{5/4}. \tag{7.10}$$

Combining these results with Theorem 6.3, we see that these estimates are also valid for reflected Brownian motions. In particular, let us now define a reflected Brownian motion $\tilde{Z}$ in the unit disc as in Theorem 6.3 (reflected on

$\partial \mathbb{U}$ with angle $2\pi/3$ "away" from $\tilde{Z}_0 = 1$). Let $\tilde{\sigma}_r$ denote its hitting time of the circle $r\partial \mathbb{U}$. Then there exist constants c_1 and c_2 such that for all $r < 1/2$,

$$c_1 r^{1/4} \leq \mathbf{P}[\tilde{Z}[0, \tilde{\sigma}_r]\text{does not disconnect 0 from } -1] \leq c_2 r^{1/4}. \qquad (7.11)$$

Similarly, (7.10) holds if one replaces K_{τ_r} by $\tilde{Z}[0, \tilde{\sigma}_r]$.

We will see in the next chapter that this also yields the corresponding estimates for (non-reflected) Brownian motions.

Bibliographical Comments

The material of this chapter is borrowed from [87], in which the reader can find more detailed proofs. It is possible to compute analogous exponents for chordal SLE. These "half-plane exponents" are determined in [86, 88].

Other important exponents are derived in [118, 92, 15]. As in this chapter, the exponents appear always as leading eigenvalues of some differential operators.

8 Brownian Exponents

8.1 Introduction

The goal of this chapter is to relate the previous computations to the exponents associated to planar Brownian motion itself (not only to reflected Brownian motion).

Suppose that a planar Brownian motion Z is started from 1. Let σ_r denote its hitting time of the circle of radius $r > 0$, and let

$$p_r := \mathbf{P}[\mathcal{D}(Z[0, \sigma_r])],$$

where $\mathcal{D}(K)$ denotes the event that K does not disconnect the origin from infinity. Note that by inversion, $p_R = p_{1/R}$ for all $R > 1$ (one can map the disk $\{|z| < R\}$ conformally on $\{|z| > 1/R\}$ by $z \mapsto 1/z$ and use conformal invariance of planar Brownian motion).

The strong Markov property and the scaling property of planar Brownian motion imply readily that for all $R, R' > 1$,

$$p_{RR'} \leq \mathbf{P}[\mathcal{D}(Z[0, \sigma_R]) \text{ and } \mathcal{D}(Z[\sigma_R, \sigma_{RR'}])] \leq p_R p_{R'}.$$

On the other hand, it is not difficult to see that

$$p_R \geq \mathbf{P}[Z[0, \sigma_R] \cap [-R, 0] = \emptyset] \geq cR^{-1/2}$$

for all $R > 1$ and some constant c. Hence, a standard subadditivity argument implies that there exists a constant $\eta \leq 1/2$ such that

$$p_R \approx R^{-\eta}$$

when $R \to \infty$, where this notation means that $\log p_R \sim -\eta \log R$. It turned out that there seems to be no direct way to determine the value of this exponent η.

Similarly, if Z^1 and Z^2 denote two independent Brownian motions started uniformly on the unit circle, then subadditivity implies the existence of a positive constant ξ such that

$$\mathbf{P}[Z^1[0, \sigma_R^1] \cap Z^2[0, \sigma_R^2] = \emptyset] \approx R^{-\xi}.$$

The exponents η and ξ are respectively called the disconnection exponent and the intersection exponent for planar Brownian motion.

8.2 Brownian Crossings

We now make some considerations that will help us relating the results on reflected Brownian motions derived in the previous chapter to the exponents η and ξ. For simplicity, we first focus on the disconnection exponent η.

Suppose that Z denotes a planar Brownian motion that is started from 1, and define the random times:

$$\sigma_r := \inf\{t > 0 \ : \ |Z_t| = r\}$$
$$\sigma_r^\# := \max\{t < \sigma_r \ : \ |Z_t| = 1\}$$
$$\sigma_r^* = \inf\{t > \sigma_r^\# \ : \ |Z_t| = 1/2\}.$$

It is a fairly standard application of the decomposition of the path $\Re(\log Z)$ into excursions away from the origin to see that

- The paths $P_r^1 := (Z_t, t \in [0, \sigma_r^\#])$ and $P_r^2 := (Z_{t+\sigma_r^\#}/Z_{\sigma_r^\#}, t \in [0, \sigma_r - \sigma_r^\#])$ are independent.
- The law of $P_r^3 := (Z_{t+\sigma_r^*}/Z_{\sigma_r^*}, t \in [0, \sigma_r - \sigma_r^*])$ is identical to the conditional law of $(Z_t, t \le \sigma_{2r})$ on the event $E_r := \{Z_{[0,\sigma_{2r}]} \subset 2\mathbb{U}\}$.

Note also that $\mathbf{P}[E_r] = \log 2 / \log r$ because $\log |Z|$ is a local martingale. We will call P_r^2 a Brownian crossing of the annulus $\mathcal{A}_r := \{1 > |z| > r\}$.

When $r' < r$, one can construct a Brownian crossing of the annulus $\mathcal{A}_{r'}$ starting from a crossing P_r^2 of the annulus $\mathcal{A}_r$ as follows: Attach to the endpoint $e_r := Z_{\sigma_r}/Z_{\sigma_r^\#}$ of P_r^2 a Brownian motion started from e_r, that is conditioned to hit the circle of radius r' before the unit circle, and stop it at that hitting time of the circle of radius r' (note that this event has probability $\log(1/r)/\log(1/r')$).

We now define the probability p_r^* that the crossing does not disconnect the origin from infinity:

$$p_r^* := \mathbf{P}[\mathcal{D}(P_r^2)].$$

Since a crossing is a subpath of a stopped Brownian motion, it follows from the a priori lower bound for p_r that $p_r^* \ge cr^{1/2}$ for some absolute constant c.

We now define for $\delta > 0$,

$$p_r^*(\delta) := \mathbf{P}[\mathcal{D}(P_r^2 \cup \mathcal{B}(1, \delta) \cup \mathcal{B}(e_r, \delta))],$$

where $\mathcal{B}(z, r)$ stands for the ball of radius r around z.

The following observations will be useful:

Lemma 8.1. *There exists $\delta > 0$ and $\varepsilon > 0$ such that for all integer n, then for at least 99% of the integers $j \in \{1, \ldots, n\}$, one has*

$$p_{r_j}^*(\delta) > \varepsilon p_{r_j}^*$$

where $r_j = 2^{-j}$.

Proof. We only sketch the main ideas of the proof. First, notice that $j \mapsto p_{r_j}^*$ is decreasing in j so that the a priori lower bound for $p_{r_j}^*$ implies that there exists ε such that for all n, then for at least 99% of the values of j in $\{1, \ldots, n-1\}$, $2\varepsilon p_{r_j}^* \le p_{r_{j+1}}^*$ (otherwise p_{r_n} would be too small). On the other hand, it is easy to see that there exists $\delta > 0$ such that

$$p^*_{r_{j+1}} \leq \varepsilon(p^*_{r_j} - p^*_{r_j}(\delta)) + p^*_{r_j}(\delta).$$

This is due to the fact that one can construct a sample of $P^2_{r_{j+1}}$ by extending the crossing $P^2_{r_j}$ into a crossing of $\{\sqrt{2} > |z| > r/\sqrt{2}\}$ by attaching conditioned Brownian motions to both ends (and then rescale this into a crossing of $\mathcal{A}_{r_{j+1}}$). And if δ is sufficiently small, then each of the attached parts disconnect the ball of radius δ around their starting point with very high probability. It therefore follows that "for 99% of the values of j",

$$p^*_{r_j}(\delta) \geq p^*_{r_{j+1}} - \varepsilon p^*_{r_j} \geq \varepsilon p^*_{r_j}.$$

$\square$

Lemma 8.2. *For all fixed δ, for some constant $c = c(\delta)$,*

$$\mathbf{P}[P^1_r \subset \mathcal{B}(1, \delta/2)] \geq \frac{c}{\log(1/r)}.$$

Proof. With positive probability, Z hits the circle of radius $1 - \delta/4$ around 0 before $\partial\mathcal{B}(1, \delta/2)$. Then, if this is the case, with probability $\log(1/(1 - \delta/4))/\log(1/r)$ it hits the circle of radius r before going back to the unit circle. $\square$

8.3 Disconnection Exponent

We now use combine these considerations with the computation of the exponents for reflected Brownian motion to prove the following result:

Theorem 8.3. *One has $\eta = 1/4$. Furthermore, there exist two constants c_1 and c_2 such that for all $R > 1$,*

$$c_1 R^{-1/4} \leq p_R \leq c_2 R^{-1/4}.$$

As we shall see later, it is important to have estimates "up-to-constants" as in this Theorem (rather than $\approx$) in order to make the link with Hausdorff dimensions.

Proof. By inversion, this is equivalent to corresponding result for small r i.e., that for all $r < 1$,

$$c_1 r^{1/4} \leq p_r \leq c_2 r^{1/4}. \tag{8.1}$$

In order to compare p_r to $\tilde{p}_r$ (this is the non-disconnection probability for reflected Brownian motion that was defined at the end of the previous chapter where we proved that it is close to $r^{1/4}$), we will in fact compare both to p^*_r.

First, one can notice using the previous lemma that

$$p_r \geq \mathbf{P}[\mathcal{D}(P^2_r \cup \mathcal{B}(1, \delta)) \cap \{P^1_r \subset \mathcal{B}(1, \delta/2)\}]$$
$$\geq p^*_r(\delta) \times \frac{c}{\log(1/r)}$$

for some constant c which is independent of $r < 1/2$. The same argument can be adapted to the reflected Brownian motion $\tilde{Z}$. Hence, "for 99% of j's",

$$p_{r_j} \geq \frac{cp^*_{r_j}}{j} \text{ and } \tilde{p}_{r_j} \geq \frac{cp^*_{r_j}}{j}$$

for some universal constant c.

On the other hand, let us now define inductively the stopping times: $\rho_0 = 0$ and for all $n \geq 0$,

$$\tau_n := \inf\{t > \rho_n : |Z_t| = 1/2\}$$
$$\rho_{n+1} := \inf\{t > \tau_n : |Z_t| = 1\}$$

the successive times of downcrossings and upcrossings between the two circles $\{|z| = 1\}$ and $\{|z| = 1/2\}$. Let N_r denote the number of upcrossings before σ_r. In other words,

$$N = N(r) := \max\{n \geq 0 : \rho_n < \sigma_r\}.$$

Note that the probability that a Brownian motion started on the circle $\{|z| = 1/2\}$ hits $\{|z| = r\}$ before the unit circle is $c_r := \log 2/\log(1/r)$, because $\log |Z|$ is a local martingale. Hence, $\mathbf{P}[N_r \geq n] = (1 - c_r)^n$. For each $n \geq 0$, the probability that $Z[\rho_n, \tau_n]$ disconnects 0 from the unit circle and does not hit the circle of radius $1/4$ is strictly positive (and independent from n). Note that if $Z[0, \sigma_r]$ does not disconnect the origin from the unit circle, then for all $n \leq N$, $Z[\rho_n, \tau_n]$ does not disconnect the origin from the unit circle, and $Z[\tau_n, \sigma_{n,r}]$ doesn't either, where

$$\sigma_{n,r} = \inf\{t > \tau_n : |Z_t| = r\}.$$

It follows that for some absolute constant $c > 0$,

$$p_r \leq \sum_{n \geq 0} (1 - c)^n (1 - c_r)^n \mathbf{P}[\mathcal{D}(Z[\tau_n, \sigma_{n,r}])]$$
$$\leq \frac{p^*_{2r}}{cc_r}$$
$$\leq \frac{\log 2}{c} \times \frac{p^*_{2r}}{\log 1/r}$$

A close inspection at the proof actually shows that the very same proof goes through if one replaces the Brownian motion Z by the reflected Brownian motion $\tilde{Z}$. Hence, for some absolute constant c',

$$p_r \leq c' \frac{p^*_{2r}}{\log(1/r)} \text{ and } \tilde{p}_r \leq c' \frac{p^*_{2r}}{\log(1/r)}.$$

Putting the pieces together, we see that "for 98% of j's",

$$\tilde{p}_{r_j} \le c_1 \frac{p^*_{2r_j}}{\log(1/r_j)} \le c_2 p_{2r_j} \le c_3 \frac{p^*_{2r_j}}{\log(1/r_j)} \le c_4 \tilde{p}_{4r_j}.$$

But we know that $r^{-1/4}\tilde{p}_r$ is bounded and bounded away from zero. It therefore follows that for some absolute constants c_1 and c_2 and at least 98% of the j's,

$$c_1 r_j^{1/4} \le p_{r_j} \le c_2 r_j^{1/4}.$$

It then remains to get rid of the last 2% of "bad" values of j. This can be done by pasting together "good" configurations that are "well-separated at the end" of the annuli $\{1 > |z| > r_{j_1}\}$ and $\{r_{j_1} > |z| > r_{j_1+j_2}\}$, where j_1 and j_2 are "good" values such that $j_1 + j_2 = j$. See for instance [91] for more details.   □

8.4 Other Exponents

The previous proofs need to be somewhat adjusted to show the corresponding result for the intersection exponent ξ (things are more complicated due to the fact that there are two Brownian motions to take care of, but no really new ideas are needed):

Theorem 8.4. *One has $\xi = 5/4$. Furthermore, there exist two constants c_1 and c_2 such that for all $R > 1$,*

$$c_1 R^{-5/4} \le \mathbf{P}[Z^1[0, \sigma_R^1] \cap Z^2[0, \sigma_R^2] = \emptyset] \le c_2 R^{-5/4}.$$

Actually, it is possible to derive the value of many other exponents. For instance, suppose that $Z^1, \ldots, Z^k, \ldots$ are independent planar Brownian motions started uniformly on the unit circle, and denote by $\sigma_R^1, \sigma_R^2, \ldots$ their respective hitting times of the circle $R\partial\mathbb{U}$, then:

Theorem 8.5. *For all $k \ge 1$, there exist constants c_1, c_2 such that for all $R > 1$,*

$$c_1 R^{-\eta_k} \le \mathbf{P}[\mathcal{D}(Z^1[0, \sigma_R^1] \cup \cdots \cup Z^k[0, \sigma_R^k])] \le c_2 R^{-\eta_k}$$

and

$$c_1 R^{-\xi_k} \le \mathbf{P}[\text{The sets } Z^1[0, \sigma_R^1], \ldots, Z^k[0, \sigma_R^k] \text{ are disjoint}] \le c_2 R^{-\xi_k},$$

where

$$\eta_k = \frac{(\sqrt{24k+1} - 1)^2 - 4}{48}$$

and

$$\xi_k = \frac{4k^2 - 1}{12}.$$

The proof of these results is however more involved. For other results and generalizations, see [86, 87, 88]. For instance, one can make sense of a continuum of exponents, or study intersection exponents for Brownian motion in a half-plane.

Let us mention that an instrumental role is also played in the definition and determination of the exponents in Theorem 8.5 by the critical exponents associated to non-intersection events in a half-space. For instance, the half-space analog of the intersection exponent ξ is:

Theorem 8.6. *If Z^1 and Z^2 are defined as before. Define*

$$q_R := \mathbf{P}[Z^1[0, \sigma_R^1] \cap Z^2[0, \sigma_R^2] = \emptyset \text{ and } Z^1[0, \sigma_R^1] \cup Z^2[0, \sigma_R^2] \subset \mathbb{H}].$$

There exist two constants c_1 and c_2 such that for all $R > 1$,

$$c_1 R^{-10/3} \leq q_R \leq c_2 R^{-10/3}.$$

There is a close relation between all these exponents (disconnection, in the whole space, in the half-space), see [96]. The critical exponents in the half-space can be determined in a similar way than the the the whole-space exponents: First one computes the "derivative" exponents associated to chordal SLE. Then, using the identification between chordal SLE_6 and reflected Brownian motion, one transfers the SLE results into Brownian motion results. For the statements and proofs of all these "half-space exponents", see [86, 88]. In order to get the value of all η_k exponents, one then uses the fact that a family of generalized exponents is analytic, see [89] for more on this.

It has also been proved (using strong approximation of simple planar random walks by Brownian motions) that these exponents describe the probabilities of the corresponding events for planar simple random walks (see [27, 37, 84, 85]). For instance, if S^1 and S^2 denote two independent simple random walks starting from neighbouring points, then

$$\mathbf{P}[S^1[0, n] \cap S^2[0, n] = \emptyset] \approx n^{-\xi/2} = n^{-5/8}$$

when $n \to \infty$ (up-to-constants hold as well). The exponent is here $\xi/2$ because we used here the parametrization in time and not in space. It is worthwhile stressing that it seems that to prove this result that seems of combinatorial nature, one has to understand and use conformal invariance of planar Brownian motion, its relation to SLE_6 as well as the properties of SLE_6.

8.5 Hausdorff Dimensions

In series of papers [76, 77, 78, 79] (before the mathematical determination of the exponents in [86, 87, 88]), Lawler showed how to use such up-to-constants estimates to estimate the Hausdorff dimension of various interesting random subsets of the planar Brownian curve in terms of the corresponding exponents.

More precisely, let $(Z_t, t \geq 0)$ denote a planar Brownian motion. Then, we say that

- The point $z = Z_t$ is a cut-point if $Z[0, t] \cap Z(t, 1] = \emptyset$.
- The point $z = Z_t$ is a boundary point if $\mathcal{D}(Z[0, 1] - z)$ i.e. if $Z[0, 1]$ does not disconnect z from infinity.
- The point $z = Z_t$ is a pioneer point if $\mathcal{D}(Z[0, t] - z)$.

Note that, loosely speaking, near $z = Z_t$, there are two independent Brownian paths starting at z: The future $Z^1 := (Z_{t+s}, s \in [0, 1 - t])$ and the past $Z^2 := (Z_{t-s}, s \in [0, t])$. Furthermore, $z = Z_t$ is a cut-point if $Z^1 \cap Z^2 = \{z\}$, z is a boundary point if $Z^1 \cup Z^2$ do not disconnect z from infinity and z is a pioneer point if Z^2 does not disconnect z from infinity. Hence, the previous theorems enable us to estimate the probability that a given point $x \in \mathbb{C}$ is in the ε-neighbourhood of a cut-point (resp. boundary point, pioneer point). Independence properties of planar Brownian paths then make it also possible to derive second moment estimates (i.e. the probability that two given points x and x' are both in the ε-neighbourhood of such points) and to obtain the following result:

Theorem 8.7.

- *The Hausdorff dimension of the set of cut-points is almost surely $2 - \xi$.*
- *The Hausdorff dimension of the set of boundary points is almost surely $2 - \eta_2$.*
- *The Hausdorff dimension of the set of pioneer points is almost surely $2 - \eta$.*

Recall that $2 - \xi = 3/4$, $2 - \eta_2 = 4/3$, $2 - \eta = 7/4$. Similar results hold for various other random subsets of the planar curve. We choose not to give the proof of this theorems in these lectures since they are more using features of planar Brownian motion rather than SLE_6, but here is a brief sketch in the case of the pioneer points.

Sketch of the proof. Let $\mathcal{P}$ denote the set of pioneer points on $Z[0, 1]$. Theorem 8.3 roughly shows that for each z, the probability that Z comes ε-close to z without disconnecting z from infinity is comparable to $\varepsilon^{1/4}$. It follows that the expectation of the number N_ε of ε-balls that are needed in order to cover $\mathcal{P}$ is comparable to (i.e. up-to-constants away from) $\varepsilon^{-2+1/4} = \varepsilon^{-7/4}$. This in fact already shows that the Hausdorff dimension of $\mathcal{P}$ can a.s. not be larger than $7/4$.

On the other hand, one has good bounds on the second moment of N_ε: This is due to the fact that for two points x and x' with $|x - x'| = r$ to be ε-close to pioneer points, then the following three events must occur before time one:

- Z reaches $\mathcal{B}(x, 2r)$ without disconnecting x
- Z crosses the annulus $\{z : \varepsilon < |z - x| < r/2\}$ without disconnecting x
- Z crosses the annulus $\{z : \varepsilon < |z - x'| < r/2\}$ without disconnecting x'.

Hence, it follows that $\mathbf{E}[N_\varepsilon^2] \leq cst \times \varepsilon^{-7/2} \leq cstE[N_\varepsilon]^2$. Standard arguments can then be used to deduce from this that with positive probability, the dimension of $\mathcal{P}$ is not smaller than 7/4. A zero-one law can finally be used to conclude that the dimension is a.s. equal to 7/4. See e.g. [82] for details.   □

Bibliographical Comments

The fact that one probably had to compute the value of the Brownian exponents via an universality argument using another model (that should be closely related to critical percolation scaling limits) first appeared in [97]. The mathematical derivation of the value of the exponents was performed in the series of papers [86, 87, 88, 89]. The properties of SLE that were later derived in [95] enable to shorten some parts of some proofs, but it seems that analyticity of the family of generalized exponents derived in [89] can not be by-passed for all exponents (for instance, it seems that it is needed to determine the exponent describing the probability that the union of three Brownian motions does not disconnect a given point). It can however be by-passed for those exponents that we have to focus on i.e., η, η_2, ξ.

Lemma 8.1 is a "separation Lemma" of the type that had been derived by Lawler in the series of papers relating the Hausdorff dimensions to the exponents [75, 76, 77, 78, 79]. The proof presented here is adapted from the proof of the analogous but more general results for the other exponents in [91]. A good reference for the relation between Brownian exponents and Hausdorff dimensions is Lawler's review paper [82]. See also, Beffara [13, 14].

Determining the Hausdorff dimensions of subsets of the SLE processes is a difficult question. Rohde-Schramm [118] have shown that the dimension of the SLE generating curve is not larger than $1 + \kappa/8$. It was conjectured to be a.s. equal to that value (for $\kappa \leq 8$). This has been proved to hold for the special values $\kappa = 8/3$ and $\kappa = 6$, making use of the locality and restriction properties (see [95], Beffara [14]). It now seems that Beffara [15] managed to prove the general conjecture.

The value of most of these exponents had been predicted/conjectured before: Duplantier-Kwon [48] had predicted the values of ξ_k using non-rigorous conformal field theory considerations, Duplantier [44] more recently used also the so-called "quantum gravity" to predict the values of all exponents. The fact that the dimension of the Brownian boundary was 4/3 was first observed visually and conjectured by Mandelbrot [107]. Before the proof of this conjecture, some rigorous bounds had been derived, for instance that the dimension of the Brownian boundary is strictly larger than 1 and strictly smaller than 3/2 (see [24, 28, 132]).

9 SLE, UST and LERW

9.1 Introduction, LERW

In the next two chapters, we will survey the rigorous results that show that for some values of κ, SLE_κ is indeed the scaling limit of discrete models. There are at present only three values of κ for which this is the case: $\kappa = 2$ is the scaling limit of LERW, $\kappa = 6$ is the scaling limit of percolation cluster interfaces, and $\kappa = 8$ is the scaling limit of the uniform spanning tree contour.

In all three cases, the convergence to SLE is derived as a consequence of three facts:

- The "Markovian" property holds in the discrete case (this is usually a trivial consequence of the definition of the microscopic model).
- Some macroscopic functionals of the model converge to conformally invariant quantities in the scaling limit (for a wide class of domains).
- One has "a priori" bounds on the regularity of the discrete paths.

Before going into more details, let us state the convergence theorem in the case of LERW that was presented in the introductory chapter: Consider γ^δ the (time-reversal of the) loop-erasure of a simple random walk in $D \cap \delta\mathbb{Z}^2$, started from 0 and stopped at the first exit time of the simply connected (say, bounded) domain D. Let γ denote a radial SLE_2 in the unit disc started uniformly on the unit circle (and aiming at 0). Let Φ denote a conformal map from $\mathbb{U}$ onto D that preserves 0. We endow the set of paths with the metric of uniform convergence modulo time-reparametrization:

$$d(\Gamma, \Gamma') = \inf_\varphi \sup_{t \geq 0} |\Gamma(t) - \Gamma'(\varphi(t))|$$

where the inf is over all increasing bijections φ from $[0, \infty)$ into itself. Then,

Theorem 9.1. *The law of γ^δ converges weakly when $\delta \to 0$ to the law of $\Phi(\gamma)$.*

Actually, one can also use the convergence result to justify the fact that SLE_2 is a simple path. Instead of giving the basic ideas of the proof of this theorem, we will focus on a closely related problem: The uniform spanning trees scaling limit.

9.2 Uniform Spanning Trees, Wilson's Algorithm

Suppose that a connected finite graph $G = (V, E)$ is given (V is the set of vertices and E is the set of edges). We say that the subgraph $T \subset E$ is a spanning tree if it contains no loop, and if it has only one connected component. We then define the uniform spanning tree as the uniform measure on the set of spanning trees. For any two fixed points a and b in G, and any spanning tree T, there exists a unique simple path in T that joins a to b (it exists because T has one connected component, it is unique because T has no loops). Hence, if T is picked according to the UST measure, this defines a random path γ from a to b. The following result had first been observed by Pemantle [113]:

Proposition 9.2 *The law of γ is that of the loop-erasure of simple random walk on G started at a and stopped at its first hitting of b.*

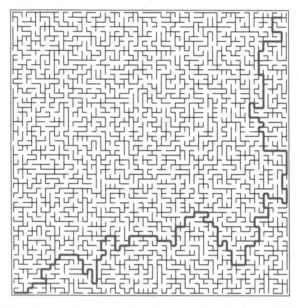

Fig. 9.1. A loop-erased walk as a subpath of the UST

This shows that LERW and UST are very closely related. Actually, it turns out that an even stronger relationship hold: Suppose that an ordering of the vertices $v_0, v_1, \ldots, v_m$ of G is given. Define inductively the sets A_m as follows: $A_0 = \{v_0\}$, and for all $j \le m$, $A_j = A_{j-1} \cup \gamma_j$ where γ_j is the loop-erasure of a random walk started from v_j and stopped at its first hitting of A_{j-1}. Clearly, in this way, A_m is a (random) tree that contains all vertices: It is a spanning tree.

Proposition 9.3 (Wilson's algorithm) *The law of A_m is the uniform spanning tree measure.*

Note that this algorithm yields a natural extension of uniform spanning trees (or forests) in infinite graphs (see e.g. [20] and the references therein for more on this subject).

Proof. One can derive this result using the explicit formulas that we derived in the introductory chapter for loop-erased random walks: Indeed, it follows readily from the definition and the symmetry of the function F that was defined there, and the fact that (since we are considering simple random walks), the transition probabilities $p(x, y)$ are simply equal to $1/d_x$ where d_x is the number of neighbours of x), that for any possible spanning tree T,

$$\mathbf{P}[A_m = T] = F(v_1, v_2, \ldots, v_m; \{v_0\}) \prod_{j=1}^{m} (1/d_{v_j}).$$

This quantity is the same for all T: The law of A_m is uniform. $\qquad\square$

Hence, if LERW has a conformally invariant scaling limit then UST also has a conformally invariant scaling limit (in a rather weak sense though, such as: for all k given fixed points, the "finite subtree that go through these points" converges in the scaling limit).

There is another way to encode planar trees that goes as follows. Suppose for instance that we are looking at a spanning tree of a bounded "simply connected" graph $G \subset \mathbb{Z}^2$. Then, one can associate to each tree the contour of the tree which is a simple closed curve living on a subset $G^{\#}$ of the lattice $(1/4 + \mathbb{Z}/2)^2$. It is easy to see that (under mild assumptions on the domain), this curve visits every point of $(1/4 + \mathbb{Z}/2)^2$ that is close to the vertices of G. If the tree is chosen according to the uniform measure on spanning trees, then the contour is chosen according to the uniform measure on space-filling simple closed curves in this graph $G^{\#}$.

Hence, it is natural to study the behaviour of this space-filling curve in the scaling limit. In order to obtain SLE (and not a closely related object that we would have to define first) it is (slightly) more convenient to consider a variant of the previously defined space-filling curve.

More precisely, suppose that a certain connected graph of $(1/4 + \mathbb{Z}/2)^2$ is given together with two distinct "boundary points" a and b. Then (for a suitable class of "admissible" graphs), one is interested in the uniform measure on simple space-filling curves η from a to b in the graph (i.e. paths from a to b that visit all vertices exactly once). An example of "admissible" graphs is given by the graph obtained from removing from $G^{\#}$ a part of a simple closed space-filling curve γ. This time, there is a one-to-one correspondence between the family of simple space-filling curve η (from a to b) and the set of spanning trees in a certain subgraph G of $\mathbb{Z}^2$ obtained by wiring one part of the boundary between a and b (i.e. by conditioning the tree to contain this part of the boundary). This is best seen on pictures, and not difficult to understand heuristically, but it is somewhat messy to formulate precisely, so we will omit the precise statements here (see e.g., [93] for more details).

Note that in this set-up, the Markovian type property for η is immediate: If one conditions on the first step $\eta(1)$ of η, then the law of $\eta(1), \ldots, \eta(n) = b$ is simply the uniform measure on the space filling curves from $\eta(1)$ to b in the remaining graph.

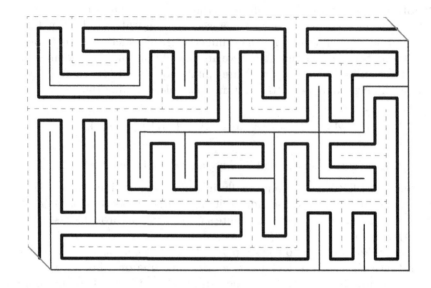

Fig. 9.2. The wired tree, the dual tree, the Peano curve.

9.3 Convergence to Chordal SLE_8

Suppose that D is a simply connected bounded planar domain with C^1 boundary and let a, b denote two distinct points on ∂D. For each δ, we associate in a "suitable approximation" of $D \cap \delta Z^2$, denoted by D_δ, and the two boundary points a_δ and b_δ close to a and b. We define η^δ, a uniformly chosen space-filling curve from a_δ to b_δ in D_δ.

Theorem 9.4. *When $\delta \to 0$, the law of η^δ converges weakly to that of a space-filling continuous path η, such that the law of $(\eta[0, t], t \geq 0)$ is (up to time-change) that of chordal SLE_8 in D from a to b.*

Some rough ideas from the proof. A first step is to obtain regularity estimates on the (discrete) random space-filling curve. This shows that the families of probability measures defining η^δ is tight in an appropriate sense and therefore has subsequential limits. These estimates have been derived in [123] (see also [5, 6]) and the basic tools are Wilson's algorithm and estimates for simple random walks. This is not easy, and we refer to [123] for details. Hence, one can work with a given decreasing sequence $\delta_n \to 0$ such that the law of η_{δ_n} converges towards that of a random curve η, and one has to show that η is in fact chordal SLE_8.

Let us first work on the discrete level. Suppose that z_δ is some discrete lattice approximation of $z \in D$ and that c_δ is some discrete lattice approximation of $c \in \partial D$ that is on the wired part of the boundary of D. Let P_1^δ denote the part of the wired boundary of D_δ which is between a_δ and c_δ, and

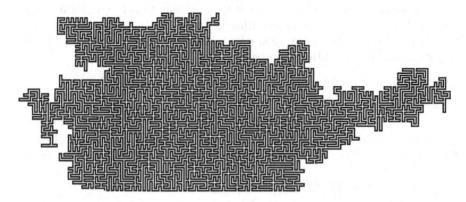

Fig. 9.3. A sample of the beginning of the Peano curve.

let P_2^δ denote the part of the wired boundary which is between c_δ and b_δ (and P_1, P_2 are defined similarly in D).

We consider the event E^δ that there exists a path in the corresponding tree that goes from z_δ to P_2^δ without touching P_1^δ. By Wilson's algorithm, we see that $\mathbf{P}[E^\delta(c, z)]$ is the probability that simple random walk on D_δ hits P_2^δ before P_1^δ. One first key-observation is that when δ goes to 0, the probability of this event can be controlled in a rather uniform way: Uniformly over some suitable choices of z, c, a, b and D, it converges towards the probability that a Brownian motion in D that is orthogonally reflected on the 'free' part of ∂D, hits P_2 before P_1. This is a conformally invariant quantity. Mapping D onto the upper half-plane by some given fixed mapping g in such a way that $g(b) = \infty$, we see that

$$\lim_{\delta \to 0} \mathbf{P}[E^\delta] := h(A, C, Z) = F\left(\frac{Z - A}{C - A}\right),$$

where

$$F(re^{i\theta}) = \frac{1}{\pi} \tan^{-1}\left(\frac{1 - r}{2\sqrt{r}\sin\theta/2}\right)$$

and $A = g(a), Z = g(z), C = g(c)$. This function F can be computed for instance by first using reflection so that this probability is the probability that (non-reflected) Brownian motion in the complex plane, started from $g(z)$ hits $[g(c), \infty)$ before $[g(a), g(z)]$, then to use the map $x \mapsto (\sqrt{x} - \sqrt{z})/(\sqrt{x} + \sqrt{z})$ from $\mathbb{C} \setminus [g(a), +\infty)$ onto the unit disk and to look at the length of the image of $[g(c), +\infty)$ on the unit circle).

At each step n, define the conformal map φ_n^δ from a continuous approximation D_n^δ of $D^\delta \setminus \eta[0, n]$ onto the upper half-plane that is characterized by $\varphi_n^\delta(x) - \varphi_0^\delta(b) = o(1)$ when $x \to b$. We then define t_n^δ to be the "size" $a(\varphi_0^\delta(\eta^\delta[0, n]))$ of $\varphi_0^\delta(\eta^\delta[0, n])$ and we put

$$A_n^\delta = \varphi_n^\delta(\eta_n), \ C_n^\delta = \varphi_n^\delta(c_\delta) \text{ and } Z_n^\delta = \varphi_n^\delta(z_\delta).$$

Suppose now that $\varepsilon > 0$ is small but fixed. If one stops the uniform Peano curve at the first step N, at which either $|A_n^\delta - A_0^\delta|$ reaches ε or t_n^δ reaches ε^2, (if c and z are not close to a), then one does not yet know whether E^δ holds or not. In fact the conditional probability is just equal to

$$\mathbf{P}[E^\delta(c_\delta, z_\delta, \eta_N, b_\delta, D_N^\delta)].$$

Hence,

$$\mathbf{E}[\mathbf{P}[E^\delta(c_\delta, z_\delta, \eta_N, b_\delta, D_N^\delta)] = \mathbf{P}[E^\delta].$$

The right-hand side is close to $h(A, C, Z)$ and the right-hand side is close to $\mathbf{E}[h(A_N^\delta, C_N^\delta, Z_N^\delta)]$ (in a uniform way as δ goes to 0). In fact, one can prove that

$$\mathbf{E}[h(A_N^\delta, C_N^\delta, Z_N^\delta)] = \mathbf{E}[h(A_0, C_0, Z_0)] + O(\varepsilon^3).$$

It turns in fact out, that the conformal map Φ_N^δ is very close to the (properly normalized) conformal map from $D \setminus \eta[0, N]$ onto $\mathbb{H}$ (i.e. removing the slit or the "tube" does not make much difference when δ is small). In particular, when ε is small (and δ very small), Loewner's equation shows that

$$(Z_N^\delta - Z_0) = \frac{2t_N^\delta}{Z_0 - A_0} + O(\varepsilon^3) \text{ and } (C_N^\delta - C_0) = \frac{2t_N^\delta}{C_0 - A_0} + O(\varepsilon^3).$$

Hence, one can Taylor-expand h in the previous estimate, so that

$$\frac{1}{2}\mathbf{E}[(A_N^\delta - A_0)^2]\partial_A^2 h(A_0, C_0, Z_0) + \mathbf{E}[A_N^\delta - A_0]\partial_A h(A_0, C_0, Z_0)$$

$$+2\mathbf{E}[t_N^\delta]\left(\frac{\partial_C h(A_0, C_0, Z_0)}{C_0 - A_0} + \frac{\partial_Z h(A_0, C_0, Z_0)}{Z_0 - A_0}\right) = O(\varepsilon^3).$$

Using the explicit expression of h as well as the fact that this holds for various values c and z yields that in fact:

$$\mathbf{E}[A_N^\delta - A_0] = O(\varepsilon^3) \text{ and } \mathbf{E}[(A_N^\delta - A_0)^2] = 8\mathbf{E}[t_N^\delta] + O(\varepsilon^3).$$

One can iterate this procedure using inductively defined stopping times $N_2, N_3, \ldots$, and one can then use this as a seed to show that it is possible to find a Brownian motion B such that A_n^δ remains close to $B_{8t_n^\delta}$, and then, after some additional work can be improved into the convergence theorem. $\square$

As the reader can see, this is only a very sketchy outline of a fairly long and technical proof. For details, see [93].

9.4 The Loop-Erased Random Walk

The strategy of the proof of Theorem 9.1 follows roughly the same lines. One has to identify a conformal invariant quantity that appears in the scaling limit of LERW and that plays the role of the probability of the events E in the case of the uniform Peano curve. The macroscopic quantities that are used are related to the mean number of visits to a given point z by the simple random walk started from 0 and conditioned to leave the domain at the same point as the LERW. See [93] for details.

Bibliographical Comments

The convergence results presented in this chapter are proved in [93], where the reader can find more details. For an introduction to LERW and UST, see for instance [104, 81]. Rick Kenyon [62, 64] had proved that LERW (and UST's) have conformally invariant features exploiting the relation between UST and dimer models (and some explicit computations). He also managed to determine directly (without using SLE_2 or SLE_8) [65, 66] the value of various critical exponents related to LERW and UST that had been conjectured by Majumdar and Duplantier [106, 43]. For instance, he showed that the expected length of a LERW from 0 to the boundary of the unit disc on the lattice $\delta\mathbb{Z}^2$ is of the order $\delta^{-5/4}$. See also, Fomin's paper [50] for another approach to some of these exponents.

In the recent preprint [71], Gady Kozma gives a completely different approach and justification to the existence of a scaling limit of LERW (that does not seem to use conformal invariance or SLE).

10 SLE and Critical Percolation

10.1 Introduction

Consider a planar "periodic" lattice such that simple random walk on that lattice converges to planar Brownian motion. For convenience, let us limit our discussion to the square lattice and to the triangular lattice. Fix $p \in [0, 1]$, and for each site of the lattice, decide that with probability p, the site is open (with probability $1 - p$, it is therefore closed), and do that independently for all sites of the lattice. One is interested in the properties of the connected components (or "clusters") of open sites. It is now classical (see e.g., [55] for an introduction to percolation) that there exists a critical value $p_c \in (0, 1)$ such that:

- If $p \leq p_c$, there exists a.s. no infinite open cluster (note that in dimension greater than 2, the non-existence of an infinite open cluster at p_c is still an open problem).
- If $p < p_c$, there exists a positive $\xi(p)$ such that when $n \to \infty$, the probability that 0 is in the same connected component than $(n, 0)$ decays exponentially fast, like $\approx \exp(-n/\xi(p))$ (the positive quantity $\xi(p)$ is called the correlation length).
- If $p > p_c$, there exists almost surely no infinite open cluster.

The value of p_c is lattice-dependent. In the case of the square lattice, it has been shown to be larger than .556 [22] (it is not expected to be any special number), while for the triangular lattice, it has been shown by Kesten and Wierman to be equal to $1/2$ (see e.g. [67]). This is not surprising because the triangular lattice has a self-matching property: It is equivalent to say that the origin is in a finite open cluster or to say that it is surrounded by a circuit (on the same lattice, this is what makes the triangular lattice so special) of closed sites. This property shows also that if $p = 1/2$ on the triangular lattice, the probability that there exists a left-to-right crossing of open sites of a square is exactly $1/2$ (otherwise, there is a top-to-bottom crossing of closed sites). Russo, Seymour and Welsh [120, 125] have shown (this is sometimes known as the RSW theory) that this in fact implies that for any fixed a and b, there exists a constant $c > 0$, such that the probability $q(aN, bN)$ of a left-to-right crossing of the $aN \times bN$ rectangle satisfies

$$1 - c > q(aN, bN) > c$$

for all large N. This strongly suggests that when $N \to \infty$, $q(aN, bN)$ converges to a limit $F(b/a)$. A renormalizing group argument (loosely speaking, the rectangle $2aN \times 2bN$ can be divided into four rectangles of size $aN \times bN$, which themselves can be divided into four rectangles etc.) also heuristically suggests that not only the crossing probabilities converge but that in some sense, the information about "macroscopic connectivity properties" should converge. Note however that things are rather subtle. Benjamini, Kalai and

Schramm [19] have for instance proved that if $A[N]$ denotes the event that there is a left-to-right crossing of a $N \times N$ square say, and if one changes the status of a fixed proportion ε of the N^2 sites and looks at the event $\tilde{A}[N]$ that there exists a left-to-right crossing for the new configuration, then the events $A[N]$ and $\tilde{A}[N]$ are asymptotically independent when $N \to \infty$. These events are "sensitive to noise". When N is large, it is not easy to "see" whether the crossing events occur or not (in the Figure 10.1, each occupied site on the triangular lattice is represented by a white hexagon).

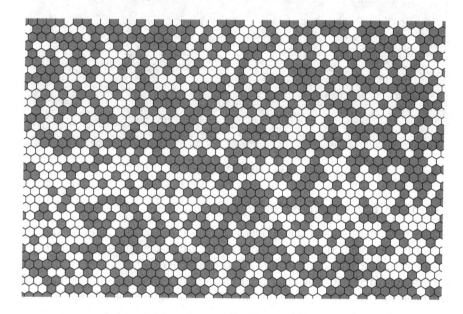

Fig. 10.1. Is there a left to right crossing of white hexagons?

In fact, the renormalization argument suggests that even though the value of p_c is lattice-dependent, on large scale, what one sees at the value p_c becomes lattice-independent. In other words, in the scaling limit, the behaviour of critical percolation should become lattice-independent (just as simple random walk converges to Brownian motion, for all "regular" lattices). Hence, the function $F(b/a)$ should be a universal function describing the crossing-probabilities of a "continuous percolation process." In fact, this continuous percolation should be scale-invariant (it is a scaling limit) as well as rotationally invariant (which would follow from lattice-independence). This leads to the stronger conjecture that it should be conformally invariant: The connections in a domain D and those in a domain D' should have the same law, modulo a conformal map from D onto D'.

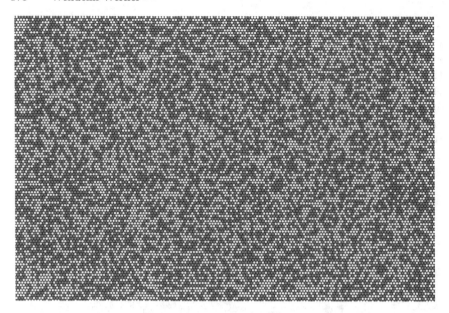

Fig. 10.2. And now?

10.2 The Cardy-Smirnov Formula

Using the conformal field theory ideas developed in [18, 30], John Cardy [31] gave an exact prediction for the function F. Extensive numerical work (e.g., [73]) did comfort these predictions. Carleson noted that Cardy's function F is closely related with the conformal maps from rectangles onto equilateral triangles, and that Cardy's prediction could be rephrased as follows:

Conjecture 10.1 (Cardy's formula). If D is conformally equivalent to the equilateral triangle OAC, and if the four boundary points a, o, c, x are respectively mapped onto $A, O, C, X \in [CA]$, then (in the scaling limit when the mesh of the lattice goes to zero), the probability that there exists a crossing in D from the part (ao) of ∂D to (cx) is equal to CX/CA.

We have seen that the SLE approach did provide a new justification to this formula. Indeed, if the percolation exploration path has a conformally invariant scaling limit, it must be one of the chordal SLEs, as argued it the first Chapter. Also, as the hitting probabilities computations in Chapter 3 show, SLE_6 is the unique SLE such that for all $X \in [CA]$, the two following probabilities are identical:

- The SLE from O to A in the equilateral triangle hits AX before XC
- The SLE from O to X in the equilateral triangle hits AX before XC

This has to hold for the scaling limit of the critical exploration process. Hence, the unique possible conformally invariant scaling limit of the critical exploration process is SLE_6. Another way to justify this is that this scaling limit has to satisfy locality and (cf. Chapter 4) that SLE_6 is the unique SLE that satisfies locality. Yet another (simpler) justification is that SLE_6 is the unique SLE for which the probability of the event corresponding to a left-right crossing of a square (or a rhombus) is $1/2$ (for an SLE starting from one corner and aiming at a neighbouring corner).

We have also seen that an SLE_6 from O to C in the equilateral triangle hits XC before AX with probability CX/CA. Also, for the discrete exploration process, the corresponding event is precisely the event that there exists a crossing from AO to CX. Hence, we get a conditional result of the following type: If the scaling limit of critical percolation exists and is conformally invariant, then the scaling limit of the exploration process is SLE_6 and Cardy's formula holds.

But in order to prove conformal invariance of critical percolation, one has to work with discrete percolation itself. In 2001, Stas Smirnov, proved that:

Theorem 10.2. *Cardy's prediction is true in the case of critical site percolation on the triangular lattice.*

In fact, Smirnov's proof is a direct proof of Cardy's formula that does not rely at all on SLE. Then, with Smirnov's result, one can show that indeed the scaling limit of the percolation exploration process is SLE_6.

Sketch of the proof. Suppose first for convenience that AOC is an equilateral triangle and that the sides of the triangle have unit length and are parallel to the axis of the triangular grid (as we will see, this has in fact no other influence on the proof than simplifying the notations). For all $\delta = 1/n$, consider critical site percolation in AOC on the triangular grid with mesh-size $1/n$. For convenience, put $\tau = \exp(2i\pi/3)$ and write $A_1 = A$, $A_\tau = A_2 = O$ and $A_{\tau^2} = A_3 = C$. For each face z of the triangular grid (i.e. for each site of the dual hexagonal lattice), let $E_1(z)$ denote the event that there exists a simple open (i.e. white) path from $A_1 A_\tau$ to $A_1 A_{\tau^2}$ that separates z from $A_\tau A_{\tau^2}$. Similarly, define the events $E_\tau(z)$ and $E_{\tau^2}(z)$ corresponding to the existence of simple open paths separating z from $A_1 A_{\tau^2}$ and $A_1 A_\tau$ respectively. Define finally for $j = 1, \tau, \tau^2$,

$$H_j(z) = H_j^\delta(z) := \mathbf{P}[E_j(z)].$$

The Russo-Seymour-Welsh theory ensures that the functions H_j^δ are uniformly "Hölder" (actually, one first has to smooth out their discontinuities for instance in a linear way keeping only the values of H_j^δ at the center of the triangles). In particular, it shows that any for any sequence $\delta_n \to 0$, the triplet of functions $(H_1^\delta, H_\tau^\delta, H_{\tau^2}^\delta)$ has a subsequential limit. Our goal is now to identify the only possible such subsequential limit.

The Russo-Seymour-Welsh estimates also show that when $z \to A_{j\tau} A_{j\tau^2}$, the functions H_j^δ go uniformly to zero, and that when $z \to A_j$, the functions H_j^δ go uniformly to one. Hence, for any subsequential limit $(H_1, H_\tau, H_{\tau^2})$, one has $H_j(z) \to 0$ when $z \to A_{j\tau} A_{j\tau^2}$, and $H_j(z) \to 1$ when $z \to A_j$.

Now comes the key-observation of combinatorial nature: Suppose that z is the center of a triangular face. Let z_1, z_2, z_3 denote the three (centers of the) neighbouring faces (with the same orientation as the triangle $A_1 A_2 A_3$) and s_1, s_2, s_3 the three corners of the face containing z chosen in such a way that s_j is the corner "opposite" to z_j. We focus on the event $E_1(z_1) \setminus E_1(z)$. This is the event that there exists three disjoint paths l_1, l_2, l_3 such that

- The two paths l_2 and l_3 are open and join the two sites s_2 and s_3 to $A_1 A_3$ and $A_1 A_2$ respectively.
- The path l_1 is closed (i.e., it consists only of closed sites), and joins s_1 to $A_2 A_3$.

One way to check whether this event holds is to start an exploration process from the corner A_3, say (leaving the open sites on the side of A_1 and the closed sites on the side of A_2). If the event $E_1(z_1) \setminus E_1(z)$ is true, then the exploration process has to go through the face z, arriving into z through the edge dual to $s_1 s_2$. In this way, one has "discovered" the simple paths l_2 and l_1 that are "closest" to A_3. Then, in the remaining (unexplored domain), there must exist a simple open path from s_3 to $A_1 A_3$. But, the conditional probability of this event is the same as that of the existence of a simple closed path from s_3 to $A_1 A_3$ (interchanging open and closed in the unexplored domain does not change the probability measure). Changing all the colors once again, shows finally that $E_1(z_1) \setminus E_1(z)$ has the same probability as the event that there exist three disjoint paths l_1, l_2, l_3 such that

- The paths l_1 and l_3 are open and join the two sites s_1, s_3 to $A_2 A_3$ and $A_1 A_2$ respectively.
- The path l_2 is closed, and joins s_2 to $A_1 A_3$.

This event is exactly $E_\tau(z_2) \setminus E_\tau(z)$. Hence, we get that,

$$\mathbf{P}[E_1(z_1) \setminus E_1(z)] = \mathbf{P}[E_\tau(z_2) \setminus E_\tau(z)] = \mathbf{P}[E_{\tau^2}(z_3) \setminus E_{\tau^2}(z)].$$

These identities can then be used to show that for any equilateral contour Γ (inside the equilateral triangle), the contour integrals of H_j^δ for $j = 1, \tau, \tau^2$ are very closely related:

$$\int_\Gamma dz H_1^\delta(z) = \int_\Gamma dz H_\tau^\delta(z)/\tau + O(\delta^\varepsilon) = \int_\Gamma dz H_{\tau^2}^\delta(z)/\tau^2 + O(\delta^\varepsilon)$$

when $\delta \to 0$ for some $\varepsilon > 0$. To see this, one has to expand the contour integrals as the sum of all properly oriented contour integrals along all small triangles inside Γ. Then, the previous identities ensure that almost all terms cancel out. The remaining "boundary" terms are controlled with the help of RSW estimates.

This result then shows that for any subsequential limit $(H_1, H_\tau, H_{\tau^2})$, the contour integrals of H_1, H_τ/τ and of H_{τ^2}/τ^2 coincide. It readily follows that the contour integrals of the functions

$$H_j + \frac{i}{\sqrt{3}}(H_{j\tau} - H_{j\tau^2})$$

for $j = 1, \tau, \tau^2$ vanish. By Morera's theorem (see e.g. [1]), this ensures that these functions are analytic. In particular, H_1 is harmonic. The boundary conditions $H_j = 0$ on $A_{j\tau}A_{j\tau^2}$ for $j = 1, \tau, \tau^2$ then ensure that $H_1 = 0$ on A_2A_3 and that the horizontal derivative of H_1 on $A_1A_3 \cup A_2A_3$ vanishes. Also, $H_1(A_1) = 1$. The only harmonic function in the equilateral triangle with these boundary conditions is the height

$$H_1(z) = \frac{d(z, BC)}{d(A, BC)}.$$

This completes the proof of the Theorem when the domain is an equilateral triangle.

If D now any simply connected domain, and $a = a_1$, $o = a_\tau$, $c = a_{\tau^2}$ are boundary points, the proof is almost identical. In its first part, the only difference is that one replaces the straight boundaries $A_jA_{j\tau}$ by approximations of the boundary of D on the triangular lattice that is between the points $a_ja_{j\tau}$. In exactly the same way, one obtains tightness and boundary estimates for the discrete functions H_j^δ. Also, the argument leading to the fact that the contour integrals on equilateral triangles of $H_j + i(H_{j\tau} - H_{j\tau^2})/\sqrt{3}$ for any subsequential limit vanish, remains unchanged. Hence, for any subsequential limit, one obtains a triplet of functions $(H_1, H_\tau, H_{\tau^2})$ such that for $j = 1, \tau, \tau^2$:

- The function $H_j + i(H_{j\tau} - H_{j\tau^2})/\sqrt{3}$ is analytic
- The function $H_j(x)$ tends to zero when x approaches the part of the boundary between $a_{j\tau}$ and $a_{j\tau^2}$.
- The function $H_j(x)$ tends to one when $x \to a_j$.

The important feature is that this problem is conformally invariant: If Φ denotes a conformal map from D onto the equilateral triangle such that $\Phi(a_j) = A_j$, and if $(H_1, H_\tau, H_{\tau^2})$ is such a triplet of functions, then the triplet $(H_1 \circ \Phi^{-1}, H_\tau \circ \Phi^{-1}, H_{\tau^2} \circ \Phi^{-1})$ solves the same problem in the equilateral triangle. In the latter case, we have seen that the unique solution is given by $H_j(x) = d(x, A_{j\tau}A_{j\tau^2})/d(A_j, A_{j\tau}A_{j\tau^2})$. Hence, the Theorem follows.   □

One should stress that this proves much more than just the asymptotic behaviour of the crossing probabilities. It yields the asymptotic probability of the events $E_j(x)$ for x inside the domain D (and not only on its boundary).

10.3 Convergence to SLE_6 and Consequences

One can use the previous result to prove that the discrete exploration process described in the introductory chapter indeed converges to chordal SLE_6.

The regularity estimates are provided by the RSW theory and the discrete Markovian property is immediate. It remains to show that some macroscopic quantities converge to a conformally invariant quantity in the scaling limit, but this is precisely what Smirnov's theorem shows. Hence, the method described in the previous chapter can be applied. Some adjustments are needed to take care of domains with rough boundary, though. In particular, one can use the a priori bounds on the probability of having 5 arms joining the vicinity of the origin to a large circle (the exponent α_5 below) derived in [69].

Exploiting this, one can therefore use the computations of critical exponents for SLE_6, to deduce asymptotic probabilities for discrete critical percolation on the triangular lattice: For instance [128, 92], let $A_n[N]$ denote the event that there exists n disjoint open clusters joining the vicinity of the origin to the circle of radius N. Then:

Theorem 10.3. *When $N \to \infty$, one has $\mathbf{P}[A_n[N]] \approx N^{-\alpha_n}$, where $\alpha_1 = 5/48$ and for all $n \geq 2$, $\alpha_n = (4n^2 - 1)/12$.*

Note that the exponents α_n for $n \geq 2$ are the same than the Brownian intersection exponents ξ_n in Chapter 8. This is not surprising because of the close relation between SLE_6 and planar Brownian motion. The exponent α_1 corresponds to the event that radial SLE_6 winds only "in one direction" around 0 (see [92].

Actually, Harry Kesten [68] had shown that the previous result (for $n = 1$ and $n = 2$) would imply the following description of the behaviour of percolation when the probability is near to the critical probability:

Theorem 10.4. *If one performs site percolation on the triangular lattice with probability p, then when $p \to 1/2+$, the probability that the origin belongs to the infinite cluster behaves like $(p - 1/2)^{5/36+o(1)}$. When $p \to 1/2-$, the correlation length explodes like $(1/2 - p)^{-4/3+o(1)}$.*

See [68, 128] for more results as well as for the proofs...

Let us conclude with the following combination of results that we have mentioned in these lectures: The following three curves are (locally) the same:

- The outer boundary of the scaling limit of a large critical percolation cluster.
- The outer boundary of a planar Brownian motion.
- The scaling limit of long self-avoiding walks, provided this scaling limit exists and is conformally invariant.

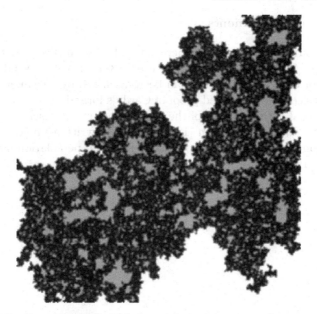

Fig. 10.3. Part of a (big) critical percolation cluster on the square lattice

Fig. 10.4. A critical percolation cluster on the triangular lattice

Bibliographical Comments

The value of critical exponents for percolation had been predicted by theoretical physicists [38, 112, 109, 110, 122, 121, 54, 33, 7]. The conformal invariance conjecture for critical percolation had been discussed by Aizenman [3, 4].

Smirnov's complete detailed proof of Cardy's formula is contained in [126, 127]. The actual detailed proof of the convergence of the discrete exploration process to SLE_6 (announced in [126]) should be written up [127] soon. For the derivation of formulas and exponents for critical percolation using SLE_6, see [124, 92, 128].

11 What Is Missing

11.1 A List of Ideas

We have listed at the end of each chapter a list of references to papers that develop ideas that are related to those presented in the corresponding chapter. One aspect of SLE that we could have spent more time on is the actual computation of critical exponents. For simplicity, we have shown how to derive the Brownian exponents using radial SLE_6, but in general (for instance to derive the Hausdorff dimension of the SLE), one might as well work with chordal SLE. Various exponents are derived in for instance in [86, 87, 88, 92, 118, 14, 15].

Before very briefly reviewing the results related to restriction properties, we would like to stress that the important ideas underlying Rohde-Schramm's [118] proof of the existence and transience of the SLE paths have not been presented in these lectures. The arguments [118] require some non-trivial background in complex geometry. In two cases, the existence and/or transience of the SLE path is especially difficult to establish: For $\kappa = 4$, because the domains generated by the SLE curve are not Hölder (see [118]). For $\kappa = 8$, the only proof uses the fact that it is the scaling limit of the discrete uniform Peano curves [93] described in Chapter 9.

One can also study geometric questions such as: Does the SLE have (local) cut points? The answer is positive if and only if $\kappa < 8$ (see [14]).

I plan to discuss the following restriction properties in forthcoming lecture notes. The main reference is the long recent paper [95].

- The full classification of the measures satisfying the restriction properties is one of the main goals of [95]. These measures form a one-dimensional family indexed by a positive real-valued parameter N, that can be interpreted as the number of Brownian excursions that the measure is equivalent to. There exist two other important ways to describe this one-dimensional family: The first one is via a variant of the $SLE_{8/3}$ process called $SLE(8/3, \rho)$. Loosely speaking, one replaces the driving Brownian motion by a Bessel process (see [95] for all this), and the obtained simple random curve describes the outer boundary of the set satisfying the restriction property. The second description goes as follows: Consider an SLE_κ with $\kappa \leq 8/3$ and add to this path a certain cloud Brownian loops (this Poisson cloud of loops is also studied in [98]). For a well-tuned density $d(\kappa)$ of the loops, one constructs the restriction measure corresponding to $N(\kappa)$ Brownian excursions. See also [40].

- This last description makes it possible to tie a link [52, 53] with representation theory, and more precisely with highest-weight representations of the Lie Algebra of polynomial vector fields on the unit circle (the number $N(\kappa)$ is the highest-weight). This is related to considerations from conformal field theory. See also [10, 11, 12] for the relation of SLE with ideas from conformal field theory.

- The $SLE(\kappa, \rho)$ processes shed also some light on the computation of the (chordal) critical exponents. It turns out that they can be understood via the absolute continuity relations between Bessel processes (following from Girsanov's Theorem); see [135].

11.2 A List of Open Problems

Here is a list of open problems. Some of these were already mentioned in the previous chapters:

Conformal Invariance of Discrete Models

So far, convergence of natural discrete models towards SLE in the scaling limit has been proved only in the two very special cases that we described in the last two chapters (LERW-UST, and critical site percolation on the triangular lattice). It is believed to hold for many other models:

- The interface for a critical FK-percolation (see e.g. [56] for an introduction to this dependent percolation model introduced by Fortuin and Kasteleyn) model for $q \leq 4$ is conjectured to converge to chordal SLE_κ. Recall that the probability of a given realization is proportional to

$$p^{\#\text{open edges}}(1-p)^{\#\text{closed edges}}q^{\#\text{connected components}}.$$

The relation between q and κ should be

$$\cos\frac{4\pi}{\kappa} = -\frac{\sqrt{q}}{2},$$

where $q \in [0,4]$ and $\kappa \in [4,8]$. Here (as in the UST case and in some sense in the percolation case), the boundary conditions have to be mixed (free on one part of the boundary, wired on the other – this influences the way of counting the connected components). See [118] for a more precise statement of this conjecture. Recall that for critical FK-percolation with parameter q on the square lattice, the self-dual point is $p = \sqrt{q}/(\sqrt{q} + 1)$ (proving that this self-dual point is the critical point is another open question, but it is not directly related to the SLE question; the question on the square grid is to prove that for this value of p, the interface converges to SLE). Here self-dual means that the law of the dual graph of an FK percolation sample is also an FK percolation sample (in the dual lattice) with the same parameters (see [56]).
Recall that when $q > 4$, the FK percolation phase transition is conjectured to be a first-order transition (i.e. there can exist an infinite open cluster at the critical probability). The critical value $q = 4$ corresponds to the special case $\kappa = 4$. Recall also (see e.g. [56]) that the correlation functions

of the critical q-Potts models are the same as those of the critical FK-percolation model. Recall also that the usual percolation is the $q = 1$ FK percolation model, and that the UST can be viewed as the $q = 0$ critical FK percolation model (see e.g. [57]). For the critical FK percolation models, the Markovian property is clearly valid in the discrete case. The missing step is therefore the proof of conformal invariance.

It is interesting (and encouraging) to note that the integer values of q correspond to the "nice" values of the angles $\alpha = \pi(1 - \kappa/4)$ of the isocele triangles for which hitting distributions are uniform (Dubédat's observations [39] mentioned at the end of Chapter 2): $\cos \alpha = \sqrt{q}/2$. For $q = 1$, it is the equilateral triangle, for $q = 2$ (Ising), it is the isocele-rectangular triangle, and for $q = 3$, $\alpha = \pi/6$.

- Among all the critical percolation interfaces that are conjectured to converge to SLE_6 (this is the special case $q = 1$ in the previous conjectures), it is worth stressing two cases, for which one has self-duality (and therefore some little hope to be able to prove something): The first one is bond-percolation on the square grid, and the second one is percolation on a Voronoi tessellation (see e.g. [21]).

- There exists a special model for which (as for the Ising model and for the uniform spanning tree model), the tools and arguments developed by Kenyon seem promising: It is the so-called double-domino path, that is conjectured to converge to the special curve SLE_4 in the scaling limit.

- Note also that the Ising model itself (on the triangular lattice) has some self-duality properties (this is due to the fact that for the Ising model, there are exactly two possible states for each site). Hence, Ising cluster interfaces (for appropriate boundary conditions, and on the triangular lattice) might converge to an SLE in the scaling limit.

- For $\kappa < 4$, the relation with discrete models from statistical physics is not so clear. One relation is via the duality conjectures that we will discuss below. The main open question is the convergence of the self-avoiding walk towards the $SLE_{8/3}$ curve. Again, the main problem is to derive its conformal invariance. See [94] for a discussion. Let us insist that basically nothing is known rigorously on the asymptotic behaviour of the self-avoiding walk. For instance, to our knowledge, it has not even been disproved that the curve becomes space-filling or a straight line in the scaling limit!

- It is likely that some discrete dynamic models can be shown to converge to SLE (but their relation to models from statistical physics is unclear). For instance, variations on the Laplacian random walk description of LERW that have some conformally invariant features built in the model should in principle converge to SLEs.

Duality

Another approach to the SLE curves when $\kappa < 4$ goes as follows: It was conjectured (based on the computation of the dimensions) that in the scaling limit, the outer boundary of an $SLE_{\kappa'}$ hull for $\kappa' > 4$ at a given time looks (locally) like an $SLE_{16/\kappa'}$ curve. Hence, the SLE_κ curves for $\kappa < 4$ correspond to the outer boundary of the scaling limit of critical FK-percolation clusters. The duality has been proved to hold in two cases: $\kappa = 2$ (because of the relation between LERW and UST that respectively converge to SLE_2 and SLE_8) and $\kappa = 8/3$ (because of the restriction property considerations that allow to describe the outer boundary of conditioned SLE_6 processes in terms of $SLE_{8/3}$ processes (see [95]). In the general case, a weak form of duality has been identified by Dubédat [40], that leads to conjecture a precise identity in law between the outer boundary of an $SLE(\kappa', \rho_{\kappa'})$ process and the $SLE(16/\kappa', \rho'_{\kappa'})$ curve for well-chosen values of ρ and ρ'.

Proving this duality relation would be one way to settle the following open problem (it is only proved when $\kappa = 6$ and $\kappa = 8$): Prove that the Hausdorff dimension of the boundary of K_t is almost surely $1 + 2/\kappa$ when K_t is the hull of an SLE_κ (chordal or radial) for $\kappa > 4$. One would then combine duality with the computation of the dimension of the SLE curves in [15]. There should however also exist a direct proof of this fact that does not rely on duality.

Reversibility

The following conjecture follows very naturally from the fact that the SLEs are believed to be scaling limit of the previously described lattice models: Suppose that $\kappa \leq 8$ is given, and consider the chordal SLE_κ curve γ from a to b in a domain D (where a and b are two boundary points). One can time-reverse γ, and view it as a curve from b to a in D. Then, the law of this time-reversal should be (modulo time-change) the law of an SLE_κ curve from b to a in D. Another equivalent way of phrasing this is that if γ is the chordal SLE path in the upper half-plane, the path $-1/\gamma$ has the same law as γ (modulo time-change).

This conjecture is very natural in terms of the lattice models, but on the other hand, it is not natural at all if one thinks of the actual definition of the SLE in terms of the Loewner chain (this is very non-reversible!). In the special cases $\kappa = 6$, $\kappa = 8$ and $\kappa = 2$, the result is a consequence of the convergence of the discrete reversible models to the SLEs. So far, the reversibility of $\kappa = 8/3$ is the only one that can be proved without reference to a reversible discrete model, and the tool here is the characterization of $SLE_{8/3}$ as the unique simple random curve that satisfies the restriction property. In all other cases, the problem is to our knowledge open. This problem does not seem as out of reach as some of those that we just discussed.

Note that (as shown to me by Oded Schramm), it is possible to show that reversibility of SLE_κ fails to be true when $\kappa > 8$. This can seem surprising; more generally, the interpretation of SLE_κ when $\kappa > 8$ in terms of models from statistical physics is not well-understood. Note that the asymptotic behaviour of SLE_κ when $\kappa \to \infty$ is studied in [16].

Quantum Gravity and Conformal Field Theory

The arguments developed in conformal field theory under the name of quantum gravity suggest that some very interesting critical phenomena also occur for systems on certain random lattices. In particular, Duplantier [44, 46, 47] showed that the value of the critical exponents in the plane (those exponents that can now be understood thanks to the SLE) can be predicted using the formula proposed by Knizhnik, Polyakov and Zamolodchikov in [70], that should relate the value of the critical exponents in the plane to the corresponding exponents on random lattices.

Recent progress has been made in the rigorous understanding of some of these random systems on these random graphs; see e.g. [9, 8, 25, 26] and the references therein. It seems that (as opposed to the rigid lattice case), the behaviour of some of these systems on random lattices might be accessible by ingenious combinatorial methods.

Note [135] that the KPZ formula seems to have a simple interpretation in terms of the ρ in the $SLE(\kappa, \rho)$ processes. Maybe the combination of the determination of the exponents for SLE, and the results on random graphs will provide in the end the rigorous justification to the KPZ relation.

More generally, the relation between SLE and conformal field theory (that has started to be investigated in [10, 11, 12, 51, 52, 53]) and with the mathematical concepts used in conformal field theory needs further understanding. It is not so clear whether this will be helpful to improve the knowledge on these critical two-dimensional systems (which was after all probably the initial motivation for the conformal field framework). One related issue is to manage to define SLE on general Riemann surfaces, see [51, 137, 42].

References

1. L.V. Ahlfors, *Complex analysis*, 3rd Ed., McGraw-Hill, New-York, 1978.
2. L.V. Ahlfors, *Conformal Invariants, Topics in Geometric Function Theory*, McGraw-Hill, New-York, 1973.
3. M. Aizenman (1996), The geometry of critical percolation and conformal invariance, Statphys19 (Xiamen, 1995), 104-120.
4. M. Aizenman (1998), Scaling limit for the incipient spanning clusters, in *Mathematics of multiscale materials*, IMA Vol. Math. Appl. **99**, Springer, New York, 1-24.
5. M. Aizenman, A. Burchard (1999), Hölder regularity and dimension bounds for random curves, Duke Math. J. **99**, 419–453.
6. M. Aizenman, A. Burchard, C.M. Newman, D.B. Wilson (1999), Scaling limits for minimal and random spanning trees in two dimensions. Random Structures Algorithms **15**, 319-367.
7. M. Aizenman, B. Duplantier, A. Aharony (1999), Path crossing exponents and the external perimeter in 2D percolation. Phys. Rev. Let. **83**, 1359-1362.
8. O. Angel (2002), Growth and Percolation on the Uniform Infinite Planar Triangulation, preprint.
9. O. Angel, O. Schramm (2003), Uniform Infinite Planar Triangulations, Comm. Math. Phys., to appear
10. M. Bauer, D. Bernard (2002), SLE_k growth processes and conformal field theories Phys. Lett. **B543**, 135-138.
11. M. Bauer, D. Bernard (2003), Conformal Field Theories of Stochastic Loewner Evolutions, Comm. Math. Phys., to appear.
12. M. Bauer, D. Bernard (2003), SLE martingales and the Virasoro algebra, preprint.
13. V. Beffara (2003), On some conformally invariant subsets of the planar Brownian curve, Ann. Inst. Henri Poincaré, to appear
14. V. Beffara (2002), Hausdorff dimensions for SLE_6, preprint.
15. V. Beffara (2002), The dimension of the SLE curves, preprint.
16. V. Beffara, in preparation
17. A.A. Belavin, A.M. Polyakov, A.B. Zamolodchikov (1984), Infinite conformal symmetry of critical fluctuations in two dimensions, J. Statist. Phys. **34**, 763–774.
18. A.A. Belavin, A.M. Polyakov, A.B. Zamolodchikov (1984), Infinite conformal symmetry in two-dimensional quantum field theory. Nuclear Phys. B **241**, 333–380.
19. I. Benjamini, G. Kalai, O. Schramm (1999), Noise sensitivity of boolean functions and applications to percolation, Publ. Sci. IHES **90**, 5-43.
20. I. Benjamini, R. Lyons, Y. Peres, O. Schramm (2001), Uniform spanning forests. Ann. Probab. **29**, 1-65.
21. I. Benjamini, O. Schramm (1998), Conformal invariance of Voronoi percolation, Comm. Math. Phys. **197**, 75-107.
22. J. van den Berg, A. Ermakov (1996), A new lower bound for the critical probability of site percolation on the square lattice, Random Structures Algorithms **8**,199-212.
23. R. van den Berg, A. Jarai (2003), The lowest crossing in 2D critical percolation, Ann. Probab., to appear.

24. C.J. Bishop, P.W. Jones, R. Pemantle, Y. Peres (1997), The dimension of the Brownian frontier is greater than 1, J. Funct. Anal. **143**, 309–336.
25. M. Bousquet-Mélou, G. Schaeffer (2002), The degree distribution in bipartite planar maps: applications to the Ising model, preprint.
26. J. Bouttier, B. Eynard, Ph. Di Francesco (2002), Combinatorics of Hard Particles on Planar Graphs, preprint.
27. K. Burdzy, G.F. Lawler (1990), Non-intersection exponents for random walk and Brownian motion. I: Existence and an invariance principle, Probab. Theor. Rel. Fields **84**, 393–410.
28. K. Burdzy, G.F. Lawler (1990), Non-intersection exponents for random walk and Brownian motion. II: Estimates and applications to a random fractal, Ann. Prob. **18**, 981-1009.
29. R. Burton, R. Pemantle (1993), Local characteristics, entropy and limit theorems for spanning trees and domino tilings via transfer-impedances, Ann. Probab. **21**, 1329–1371.
30. J.L. Cardy (1984), Conformal invariance and surface critical behavior, Nucl. Phys. B **240** (FS12), 514–532.
31. J.L. Cardy (1992), Critical percolation in finite geometries, J. Phys. A, **25** L201–L206.
32. J.L. Cardy, *Scaling and renormalization in statistical physics*, Cambridge Lecture Notes in Physics **5**, Cambridge University Press, 1996.
33. J.L. Cardy (1998), The number of incipient spanning clusters in two-dimensional percolation, J. Phys. A **31**, L105.
34. J.L. Cardy (2001), Lectures on Conformal Invariance and Percolation, Lectures delivered at Chuo University, Tokyo, preprint.
35. L. Carleson, N.G. Makarov (2001), Aggregation in the plane and Loewner's equation, Comm. Math. Phys. **216**, 583-607.
36. L. Carleson, N.G. Makarov (2002), Laplacian path models, preprint
37. M. Cranston, T. Mountford (1991), An extension of a result by Burdzy and Lawler, Probab. Th. Relat. Fields **89**, 487–502.
38. M.P.M. Den Nijs (1979), A relation between the temperature exponents of the eight-vertex and the q-state Potts model, J. Phys. A **12**, 1857-1868.
39. J. Dubédat (2003), SLE and triangles, El. Comm. Probab. **8**, 28-42.
40. J. Dubédat (2003), $SLE(\kappa, \rho)$ martingales and duality, preprint.
41. J. Dubédat (2003), Reflected planar Brownian motion, intertwining relations and crossing probabilities, preprint.
42. J. Dubédat (2003), Critical percolation in annuli and SLE_6, Comm. Math. Phys., to appear.
43. B. Duplantier (1992), Loop-erased self-avoiding walks in two dimensions: exact critical exponents and winding numbers, Physica A **191**, 516–522.
44. B. Duplantier (1998), Random walks and quantum gravity in two dimensions, Phys. Rev. Lett. **81**, 5489–5492.
45. B. Duplantier (1999), Harmonic measure exponents for two-dimensional percolation, Phys. Rev. Lett. **82**, 3940-3943.
46. B. Duplantier (2000), Conformally invariant fractals and potential theory, Phys. Rev. Lett. **84**, 1363-1367.
47. B. Duplantier (2003), Conformal Fractal Geometry and Boundary Quantum Gravity, preprint
48. B. Duplantier, K.-H. Kwon (1988), Conformal invariance and intersection of random walks, Phys. Rev. Let. **61**, 2514–2517.

49. P.L. Duren, *Univalent functions*, Springer, 1983.
50. S. Fomin (2001), Loop-erased walks and total positivity, Trans. Amer. Math. Soc. **353**, 3563–3583.
51. R. Friedrich, J. Kalkkinen (2003), On Conformal Field Theory and Stochastic Loewner Evolution, preprint.
52. R. Friedrich, W. Werner (2002), Conformal fields, restriction properties, degenerate representations and SLE, C.R. Ac. Sci. Paris Ser. I Math **335**, 947-952.
53. R. Friedrich, W. Werner (2003), Conformal restriction, highest-weight representations and SLE, Comm. Math. Phys., to appear.
54. T. Grossman, A. Aharony (1987), Accessible external perimeters of percolation clusters, J.Physics A **20**, L1193-L1201
55. G.R. Grimmett, *Percolation*, Springer, New-York, 1989.
56. G.R. Grimmett (1997), Percolation and disordered systems, Ecole d'été de Probabilités de St-Flour XXVI, L.N. Math. **1665**, 153-300
57. O. Häggström (1995), Random-cluster Measures and Uniform Spanning Trees, Stoch. Proc. Appl. **59**, 267-275
58. W.K. Hayman, *Multivalent functions*, CUP, 1994 (second edition).
59. N. Ikeda and S. Watanabe, *Stochastic Differential Equations and Diffusion Processes*, Second edition, North-Holland, 1989.
60. T. Kennedy (2002), A faster implementation of the pivot algorithm for self-avoiding walks, J. Stat. Phys. **106**, 407-429.
61. T. Kennedy (2002), Monte Carlo Tests of Stochastic Loewner Evolution Predictions for the 2D Self-Avoiding Walk, Phys. Rev. Lett. **88**, 130601.
62. R. Kenyon (1997), Local statistics of lattice dimers, Ann. Inst. Henri Poincaré **33**, 591-618.
63. R. Kenyon (1999), Dimères et arbres couvrants, in *Mathématique et Physique*, SMF Journ. Annu., 1-14.
64. R. Kenyon (2000), Conformal invariance of domino tiling, Ann. Probab. **28**, 759-785.
65. R. Kenyon (2000), The asymptotic determinant of the discrete Laplacian, Acta Math. **185**, 239-286.
66. R. Kenyon (2000), Long-range properties of spanning trees in $\mathbb{Z}^2$, J. Math. Phys. **41** 1338–1363.
67. H. Kesten, *Percolation theory for mathematicians,* Birhäuser, Boston, 1982.
68. H. Kesten (1987), Scaling relations for 2D-percolation, Comm. Math. Phys. **109**, 109-156.
69. H. Kesten, V. Sidoravicius, Yu. Zhang (2001), Percolation of Arbitrary words on the Close-Packed Graph of $\mathbb{Z}^2$, Electr. J. Prob. **6**, paper no. 4.
70. V.G. Knizhnik, A.M. Polyakov, A.B. Zamolodchikov (1988), Fractal structure of 2-D quantum gravity, Mod. Phys. Lett. **A3**, 819.
71. G. Kozma (2002), Scaling limit of loop erased random walk - a naive approach, preprint.
72. P.P. Kufarev (1947), A remark on integrals of the Loewner equation, Dokl. Akad. Nauk SSSR **57**, 655-656.
73. R. Langlands, Y. Pouillot, Y. Saint-Aubin (1994), Conformal invariance in two-dimensional percolation, Bull. A.M.S. **30**, 1–61.
74. G.F. Lawler (1980), A self-avoiding random walk, Duke Math. J. **47**, 655-694.
75. G.F. Lawler, *Intersections of Random Walks,* Birkhäuser, Boston, 1991.
76. G.F. Lawler (1995), Nonintersecting planar Brownian motions, Mathematical Physics Electronic Journal **1**, paper no.1.

77. G.F. Lawler (1996), Hausdorff dimension of cut points for Brownian motion, Electron. J. Probab. **1**, paper no.2.
78. G.F. Lawler (1996), The dimension of the frontier of planar Brownian motion, Electron. Comm. Prob. **1**, paper no.5.
79. G.F. Lawler (1997), The frontier of a Brownian path is multifractal, preprint.
80. G.F. Lawler (1998), Strict concavity of the intersection exponent for Brownian motion in two and three dimensions, Mathematical Physics Electronic Journal **5**, paper no. 5.
81. G.F. Lawler (1999), Loop-erased random walk, in *Perplexing problems in Probability*, Prog. Prob. **44**, Birkhäuser, 197-217.
82. G.F. Lawler (1999), Geometric and fractal properties of Brownian motion and random walk paths in two and three dimensions, Bolyai Mathematical Society Studies, **9**, 219-258.
83. G.F. Lawler (2001), An introduction to the stochastic Loewner evolution, preprint.
84. G.F. Lawler, E.E. Puckette (1997), The disconnection exponent for simple random walk, Israel J. Math. **99**, 109-122.
85. G.F. Lawler, E.E. Puckette (2000), The intersection exponent for simple random walk, Combin. Probab. Comput. **9**, 441-464.
86. G.F. Lawler, O. Schramm, W. Werner (2001), Values of Brownian intersection exponents I: Half-plane exponents, Acta Mathematica **187**, 237-273.
87. G.F. Lawler, O. Schramm, W. Werner (2001), Values of Brownian intersection exponents II: Plane exponents, Acta Mathematica **187**, 275-308.
88. G.F. Lawler, O. Schramm, W. Werner (2002), Values of Brownian intersection exponents III: Two sided exponents, Ann. Inst. Henri Poincaré **38**, 109-123.
89. G.F. Lawler, O. Schramm, W. Werner (2002), Analyticity of planar Brownian intersection exponents, Acta Mathematica **189**, 179-201.
90. G.F. Lawler, O. Schramm, W. Werner (2001), The dimension of the planar Brownian frontier is 4/3, Math. Res. Lett. **8**, 401-411.
91. G.F. Lawler, O. Schramm, W. Werner (2001), Sharp estimates for Brownian non-intersection probabilities, in: *In and Out of Equilbrium*, V. Sidoravicius Ed., Prog. Probab. **51**, Birkhäuser, 113-131.
92. G.F. Lawler, O. Schramm, W. Werner (2002), One-arm exponent for critical 2D percolation, Electronic J. Probab. **7**, paper no.2.
93. G.F. Lawler, O. Schramm, W. Werner (2001), Conformal invariance of planar loop-erased random walks and uniform spanning trees, Ann. Probab., to appear.
94. G.F. Lawler, O. Schramm, W. Werner (2002), On the scaling limit of planar self-avoiding walks, AMS Proc. Symp. Math., Volume in honor of B.B. Mandelbrot, to appear.
95. G.F. Lawler, O. Schramm, W. Werner (2003), Conformal restriction properties. The chordal case, J. Amer. Math. Soc **16**, 917-955.
96. G.F. Lawler, W. Werner (1999), Intersection exponents for planar Brownian motion, Ann. Probab. **27**, 1601-1642.
97. G.F. Lawler, W. Werner (2000), Universality for conformally invariant intersection exponents, J. Europ. Math. Soc. **2**, 291-328.
98. G.F. Lawler, W. Werner (2003), The Brownian loop-soup, Probab. Th. Rel. Fields, to appear.
99. N.N. Lebedev, *Special Functions and their Applications*, transl. from russian, Dover, 1972.

100. J.F. Le Gall (1992), Some properties of planar Brownian motion, Ecole d'été de Probabilités de St-Flour XX, L.N. Math. **1527**, 111-235.
101. O. Lehto, K.I. Virtanen, *Quasiconformal mappings in the plane*, second edition, translated from German, Springer, New York, 1973.
102. P. Lévy, *Processus Stochastiques et Mouvement Brownien*, Gauthier-Villars, Paris, 1948.
103. K. Löwner (1923), Untersuchungen über schlichte konforme Abbildungen des Einheitskreises I., Math. Ann. **89**, 103–121.
104. R. Lyons (1998), A bird's-eye view of uniform spanning trees and forests, in *Microsurveys in Discrete Probability*, D. Aldous and J. Propp eds., Amer. Math. Soc., Providence, 135–162.
105. N. Madras, G. Slade, *The Self-Avoiding Walk*, Birkhäuser, 1993.
106. S.N. Majumdar (1992), Exact fractal dimension of the loop-erased random walk in two dimensions, Phys. Rev. Lett. **68**, 2329–2331.
107. B.B. Mandelbrot, *The Fractal Geometry of Nature*, Freeman, 1982.
108. D.E. Marshall, S. Rohde (2001), The Loewner differential equation and slit mappings, preprint.
109. B. Nienhuis, E.K. Riedel, M. Schick (1980), Magnetic exponents of the two-dimensional q-states Potts model, J. Phys A **13**, L. 189-192.
110. B. Nienhuis (1984), Coulomb gas description of 2-D critical behaviour, J. Stat. Phys. **34**, 731-761.
111. B. Nienhuis (1987), Coulomb gas formulation of two-dimensional phase transitions, in *Phase transitions and critical phenomena* **11**, Academic Press, 1–53.
112. R.P. Pearson (1980), Conjecture for the extended Potts model magnetic eigenvalue, Phys. Rev. B **22**, 2579-2580.
113. R. Pemantle (1991), Choosing a spanning tree for the integer lattice uniformly, Ann. Probab. **19**, 1559-1574.
114. A.M. Polyakov (1974), A non-Hamiltonian approach to conformal field theory, Sov. Phys. JETP **39**, 10-18.
115. C. Pommerenke (1966), On the Löwner differential equation, Michigan Math. J. **13**, 435–443.
116. C. Pommerenke, *Boundary Behaviour of Conformal Maps*, Springer-Verlag, 1992.
117. D. Revuz, M. Yor, *Continuous Martingales and Brownian Motion*, Springer-Verlag, 1991.
118. S. Rohde, O. Schramm (2003), Basic properties of SLE, Ann. Math., to appear.
119. W. Rudin, *Real and Complex Analysis*, Third Ed., McGraw-Hill, 1987.
120. L. Russo (1978), A note on percolation, Z. Wahrscheinlichkeitsth. verw. Geb. **56**, 229-237.
121. H. Saleur, B. Duplantier (1987), Exact determination of the percolation hull exponent in two dimensions, Phys. Rev. Lett. **58**, 2325.
122. B. Sapoval, M. Rosso, J. F. Gouyet (1985), The fractal nature of a diffusion front and the relation to percolation, J. Physique Lett. **46**, L149-L156
123. O. Schramm (2000), Scaling limits of loop-erased random walks and uniform spanning trees, Israel J. Math. **118**, 221–288.
124. O. Schramm (2001), A percolation formula, Electr. Comm. Probab. **6**, 115-120.
125. P.D. Seymour, D.J.A. Welsh (1978), Percolation probabilities on the square lattice, in *Advances in Graph Theory*, ann. Discr. Math. **3**, North-Holland, 227-245.

126. S. Smirnov (2001), Critical percolation in the plane: conformal invariance, Cardy's formula, scaling limits, C. R. Acad. Sci. Paris Sér. I Math. **333**, 239–24

127. S. Smirnov, in preparation.

128. S. Smirnov, W. Werner (2001), Critical exponents for two-dimensional percolation, Math. Res. Lett. **8**, 729-744.

129. B. Virag (2003), Brownian beads, Probab. Th. Rel. Fields, to appear.

130. S.R.S. Varadhan, R.J. Williams (1985), Brownian motion in a wedge with oblique reflection. Comm. Pure Appl. Math. **38**, 405–443.

131. W. Werner (1994), Sur la forme des composantes connexes du complémentaire de la courbe brownienne plane, Probab. Theory Related Fields **98**, 307–337.

132. W. Werner (1996), Bounds for disconnection exponents, Electr. Comm. Probab. **1**, 19-28.

133. W. Werner (1997), Asymptotic behaviour of disconnection and non-intersection exponents, Probab. Theory Related Fields **108**, 131-152.

134. W. Werner (2001), Critical exponents, conformal invariance and planar Brownian motion, in *Proceedings of the 4th ECM Barcelona 2000*, Prog. Math. **202**, Birkhäuser, 87-103.

135. W. Werner (2003), Girsanov's theorem for $SLE(\kappa, \rho)$ processes, intersection exponents and hiding exponents, Ann. Fac. Sci. Toulouse, to appear.

136. D.B. Wilson (1996), Generating random spanning trees more quickly than the cover time, Proceedings of the Twenty-eighth Annual ACM Symposium on the Theory of Computing (Philadelphia, PA, 1996), 296–303.

137. D. Zhan (2003), preprint.

List of Participants

ABRAHAM Romain	Univ. René Descartes, Paris, F
ALILI Larbi	ETH Zurich, Switzerland
ATTOUCH Mohamed Kadi	Univ. Djillali Liabes, Sidi Bel Abbès, Algérie
BEFFARA Vincent	Univ. Paris-Sud, Orsay, F
BELHADJI Lamia	Univ. Mostaganem, Algérie
BERESTYCKI Julien	Univ. Pierre et Marie Curie, Paris, F
BERTOIN Jean	Univ. Pierre et Marie Curie, Paris, F
BLACHE Fabrice	Univ. Blaise Pascal, Clermont-Ferrand, F
CABALLERO Maria-Emilia	UNAM, Mexico D.F., Mexico
CALKA Pierre	Univ. Claude Bernard, Lyon, F
CAMPANINO Massimo	Univ. Bologna, Italia
CAMPI Luciano	Univ. Pierre et Marie Curie, Paris, F
CAPITAINE Mireille	CNRS, Univ. Paul Sabatier, Toulouse, F
CARMONA Philippe	Univ. Paul Sabatier, Toulouse, F
CHASSAING Philippe	Institut Elie Cartan, Nancy, F
CHAUMONT Loïc	Univ. Pierre et Marie Curie, Paris, F
CHELIOTIS Dimitrios	Stanford Univ., USA
CHERIDITO Patrick	ETH Zurich, Switzerland
COUTIN Laure	Univ. Paul Sabatier, Toulouse, F
DAVIAUD Olivier	Stanford Univ., USA
DHERSIN Jean-Stéphane	Univ. René Descartes, Paris, F
DONATI-MARTIN Catherine	CNRS, Univ. Paul Sabatier, Toulouse, F
DOUMERC Yan	Univ. Paul Sabatier, Toulouse, F
DUBEDAT Julien	Ecole Normale Supérieure, Paris, F
DURRINGER Clément	Univ. Paul Sabatier, Toulouse, F
ENRIQUEZ Nathanaël	Univ. Pierre et Marie Curie, Paris, F
FERRALIS Marc	Univ. Pierre et Marie Curie, Paris, F
FRIEDRICH Roland	Univ. Paris-Sud, Orsay, F
FUSCHINI Serena	Univ. Bologna, Italia
GHERIBALLAH Abdelkader	Univ. Djillali Liabes, Sidi Bel Abbès, Algérie
GOLDSCHMIDT Christina	Univ. Cambridge, UK

GREENWOOD Priscilla	Arizona State Univ., Tempe, USA
GRORUD Axel	Univ. Provence, Marseille, F
HAAS Bénédicte	Univ. Pierre et Marie Curie, Paris, F
HERBIN Erick	Dassault Aviation, Saint-Cloud, F
HOLROYD Alexander	Univ. California, Los Angeles, USA
HU Yueyun	Univ. Pierre et Marie Curie, Paris, F
KASPI Haya	Technion, Israel
KOURKOVA Irina	Univ. Pierre et Marie Curie, Paris, F
KUPPER Michael	ETH Zurich, Switzerland
LE GALL Jean-Franċois	Ecole Normale Supérieure, Paris, F
LE JAN Yves	Univ. Paris-Sud, Orsay, F
LEURIDAN Christophe	Institut Fourier, Grenoble, F
LEVY Thierry	CNRS, IRMA, Strasbourg, F
LORANG Gerard	Centre Universitaire de Luxembourg
MAIDA Mylène	Ecole Normale Supérieure, Lyon, F
MANSUY Roger	Univ. Pierre et Marie Curie, Paris, F
MARCHAL Philippe	CNRS, Ecole Normale Supérieure, Paris, F
MARTIN-LOF Anders	Univ. Stockholm, Sweden
MATHIEU Pierre	Univ. Provence, Marseille, F
MEJANE Olivier	Univ. Paul Sabatier, Toulouse, F
MYTNIK Leonid	Technion, Israel
NIEDERHAUSEN Meike	Purdue Univ., West Lafayette, USA
NIKEGHBALI Ashkan	Univ. Pierre et Marie Curie, Paris, F
NUALART David	Univ. Barcelona, Spain
PARVIAINEN Robert	Uppsala Univ., Sweden
PECCATI Giovanni	Univ. Pierre et Marie Curie, Paris, F
PICARD Jean	Univ. Blaise Pascal, Clermont-Ferrand, F
QUER Lluís	Univ. Barcelona, Spain
RIVERO Victor	Univ. Pierre et Marie Curie, Paris, F
RIVIERE Olivier	Univ. René Descartes, Paris, F
ROMIK Dan	Univ. Pierre et Marie Curie, Paris, F
ROUAULT Alain	Univ. Versailles, F
ROUX Daniel	Univ. Blaise Pascal, Clermont-Ferrand, F
SABOT Christophe	CNRS, Univ. Pierre et Marie Curie, Paris, F
SAINT LOUBERT BIE Erwan	Univ. Blaise Pascal, Clermont-Ferrand, F
SAVONA Catherine	Univ. Blaise Pascal, Clermont-Ferrand, F
SCHMITZ Tom	ETH Zurich, Switzerland
SERLET Laurent	Univ. René Descartes, Paris, F
SKOLIMOWSKA Magdalena	Univ. Wroclaw, Poland
SZEKELY Balazs	Budapest Univ. Technol. and Econ., Hungary
TAKAOKA Koichiro	Hitotsubashi Univ., Tokyo, Japan
VALKO Benedek	Technical Univ. Budapest, Hungary
WINKEL Matthias	Univ. Oxford, UK
YASSAI Sadr	Univ. Pierre et Marie Curie, Paris, F
YOR Marc	Univ. Pierre et Marie Curie, Paris, F

List of Short Lectures

Romain ABRAHAM, Représentation probabiliste des solutions de $\Delta u = 4u^2$ dans un domaine avec condition de Neumann au bord.

Vincent BEFFARA, The dimension of the SLE_k curve.

Julien BERESTYCKI, Fast and slow points in a fragmentation.

Jean BERTOIN, Sur les petites masses dans un processus de fragmentation.

Massimo CAMPANINO, Ornstein-Zernike theory for the finite range Ising models above T_c.

Pierre CALKA, The distribution of the number of sides of the typical Poisson-Voronoi cell.

Philippe CHASSAING, Random planar maps and Brownian snake.

Loïc CHAUMONT, Sur une identité de fluctuation pour les marches aléatoires.

Patrick CHERIDITO, Moving average representation of Gaussian processes and the semimartingale property.

Yan DOUMERC, Combinatorial representations of eigenvalues of random Gaussian matrices.

Nathanaël ENRIQUEZ, Correlated random walks and their continuous time analog.

Christina GOLDSCHMIDT, Essential edges in Poisson random hypergraphs.

Bénédicte HAAS, Perte de masse dans des systèmes de fragmentation.

Erick HERBIN, Mouvements browniens multifractionnaires indexés par $\mathbb{R}_+^N$.

Alexander HOLROYD, Bootstrap percolation and $\pi^2/18$.

Haya KASPI, Lenses in skew Brownian motion.

Irina KOURKOVA, Derrida's generalised random energy model of spin glasses: a rigorous analysis.

Christophe LEURIDAN, Filtration d'une marche aléatoire stationnaire sur le cercle.

Thierry LEVY, Yang-Mills measure: a random geometry on surfaces.

Philippe MARCHAL, The simple random walk and the Chinese restaurant.

Olivier MEJANE, Upper bound of a volume exponent for directed polymers in random environment.

Leonid MYTNIK, Regularity and irregularity of $(1+\beta)$-stable super-Brownian motion.

David NUALART, Stochastic calculus with respect to the fractional Brownian motion.

Robert PARVIAINEN, Ordering bond percolation critical probabilities on Archimedean and Laves lattices.

Giovanni PECCATI, Multiple integral representation for functionals of Dirichlet processes.

Lluís QUER, Absolute continuity of the law of the solution to the three-dimensional stochastic wave equation.

Victor RIVERO, Sur des ensembles aléatoires associés aux maxima locaux d'un processus de Poisson ponctuel.

Dan ROMIK, The hook walk on continual Young diagrams.

Laurent SERLET, Poisson snake and self-similar fragmentation.

Koichiro TAKAOKA, On Kamazaki's criterion for continuous exponential martingales.

Benedek VALKO, Perturbing the hydrodynamic limit.

Matthias WINKEL, Subordination in the wide sense of Lévy processes.

Marc YOR, q-calcul, fonctionnelles exponentielles du processus de Poisson, et une solution au problème des moments de la loi log-normale.